The Arboriculturalist's Companion
A Guide to the Care of Trees

Second Edition

N. D. G. James

BLACKWELL
Publishers

Copyright © N. D. G. James 1972; 1990

First published 1972
Second edition 1990
Reprinted 1993, 1994, 1996, 1997

Blackwell Publishers Ltd
108 Cowley Road
Oxford OX4 1JF, UK

Blackwell Publishers Inc
Commerce Place, 350 Main Street,
Malden, Massachusetts 02148, USA

British Library Cataloguing in Publication Data
A CIP catalogue record for this book is available from the British Library

Library of Congress Cataloging in Publication Data
James, N. D. G.
The arboriculturalist's companion: a guide to the care of trees/
N. D. G. James – 2nd ed.
p.   cm.
ISBN   0–631–16774–9 (pbk)
1. Arboriculture. 2. Arboriculture – Great Britain. I. Title.
SB435.J35   1990   89–36953
635.9'77—dc20   CIP

Typeset in 10 on 11 pt Century Old Style
by Joshua Associates Ltd, Oxford
Printed and bound in Great Britain
by Athenæum Press Ltd, Gateshead, Tyne & Wear

This book is printed on acid-free paper

To the Memory of
A. D. C. Le Sueur, OBE, FRICS
A Great Arboriculturalist

Contents

CHAPTER I
Introduction to Arboriculture

CHAPTER II
The Planting and Care of Amenity Trees

CHAPTER III
Transplanting Large Trees

CHAPTER X
Trees and Urban Development

CHAPTER XI
Planting Industrial Waste

CHAPTER XII
Hedges, Screens and Shelter Belts

CHAPTER XIII
Hedgerow Trees

CHAPTER XIV
Avenues, Parks and Amenity Woods

CHAPTER XV
Trees and the Law

CHAPTER XVI
Felling Licences

CHAPTER XVII
Tree Preservation Orders

CHAPTER XVIII
Records and Labelling

A Records

B Labelling

CHAPTER XIX
The Forestry Commission

CHAPTER XX
Arboricultural Education, Training and Research

CHAPTER XXI
Organizations Concerned with Arboriculture

CHAPTER XXII
Botanic Gardens, Arboreta and Pineta

CHAPTER XXIII
Books, Manuals and Periodicals

CHAPTER XXIV
Trees for Urban Areas

Foreword

By J. A. SPENCER, MA, FICFor.
*President of the Royal Forestry Society of England,
Wales and Northern Ireland*

The Royal Forestry Society of England, Wales and Northern Ireland has had a long connection with arboriculture, indeed it started life as the English Arboricultural Society, and arboriculture is still included equally with forestry in the Society's objects. These are concerned in both cases with the 'advancement and dissemination of knowledge and practice', the 'advancement of education' and the promotion and publication of research, and they are just what Mr N. D. G. James has set out to achieve in the new edition of this invaluable book.

Arboriculture has become increasingly important as a profession in its own right, and the Society recognized this as long ago as 1958 when it introduced the Certificate of Arboriculture, followed in 1959 by the Diploma in Arboriculture. At the same time public awareness of trees in the landscape – in both town and countryside – has grown considerably in recent years.

It is now more important than ever that our trees are cared for with knowledge, skill and understanding. Since it was first published in 1972, *The Arboriculturalist's Companion* has made itself an indispensable handbook, and I commend it to our members and to all those who are concerned in any way with the welfare of our trees.

Preface to the Second Edition

Seldom can there have been a greater need for trees than at the present time. Building and industrial development, with the accompanying new roads served by additional motorways, all create a demand for more trees and their subsequent care and maintenance. At the same time, more information and advice is being sought on such matters.

This book is intended as a concise guide to arboriculture. It is not, nor is intended to be, an all-embracing work on the subject. I have, however, indicated at points throughout the book where additional information can be found by those who require coverage in greater detail or depth.

N. D. G. J.

BLAKEMORE HOUSE
KERSBROOK
BUDLEIGH SALTERTON
DEVON
April 1989

Acknowledgements for the
Second Edition

I wish to thank the following for the help that they have given me, for information that they have provided or for permission to include extracts from other publications: The Secretary, The Arboricultural Association; The British Standards Institution for permission to use information in British Standards; The Administrative Secretary, University Botanic Garden, Cambridge; The Forestry Commission; The Secretary, Forestry Safety Council; Mr E. H. M. Harris, lately Director, The Royal Forestry Society of England, Wales and Northern Ireland; The Superintendent, Oxford Botanic Garden; Dr E. J. Parker, Technical Publications Officer, Forestry Commission; Mr D. Patch, The Arboricultural Advisory and Information Service; The Principals of the Colleges referred to in Chapter XX; The Regius Keeper, The Royal Botanic Garden, Edinburgh; The Director, The Royal Botanic Gardens, Kew; The Secretary, The Royal Horticultural Society; The Director, The Royal Horticultural Society Garden, Wisley; The Secretary, The Royal Scottish Forestry Society; The Secretary, The Scottish Arboricultural Society; Dr A. L. Shigo, Shigo and Trees, Associates and Mr. P. L. Wilkins, Department of Leisure and Tourist Services, Bath City Council.

CHAPTER I

Introduction to Arboriculture

1 Definition of arboriculture

(*a*) The word 'arboriculture' was originally used to cover the subject of forestry as well as that of specimen and ornamental trees. In 1868 John Grigor wrote a book entitled *Arboriculture* which had as its sub-title 'A practical treatise on raising and managing forest trees'. Furthermore it was not until 1931 that the name of the Royal English Arboricultural Society was changed to the Royal English Forestry Society. Thus although arboriculture is defined in *The Concise Oxford Dictionary* as the 'cultivation of trees and shrubs', it has now come to have a more specialized meaning in order to differentiate it from silviculture.

(*b*) Arboriculture may be defined therefore as follows:
The cultivation of trees in order to produce individual specimens of the greatest ornament and beauty, to provide shelter, to produce fruit or, for any primary purpose *other* than the production of timber as such.

(*c*) Silviculture on the other hand is:
The cultivation of trees, usually in woods and forests, primarily for the production of timber.

2 Trees and amenity

(*a*) *The amenity aspect of trees*
(i) Trees can add to the amenity of a site, area or district in numerous ways. Not only can they add to the beauty of the landscape but they can provide shelter and constitute an aid to botanical knowledge.
(ii) The following are some of the purposes for which trees may be grown, other than for timber:
(1) *Ornament and beauty*
Either as individuals, in groups or as woods.

1

(2) *Shelter*
By providing a shelter against the wind or a screen to obscure an object or view.

(3) *Education*
As a means for providing botanical knowledge through the medium of arboreta, pineta and botanical gardens.

(4) *Fruit production*
Whether commercially or in private gardens.

(*b*) *Points to consider in selecting trees*
In selecting trees for a particular purpose or use, the following should be borne in mind:

(i) *Size*
The ultimate height and spread of a tree will depend largely on the species or variety. Too often trees are planted in a restricted space without consideration being given to the size which they will ultimately attain.

(ii) *Shape*
In addition to the actual size to which a tree will grow, the question of shape may well be of considerable importance. Columnar or fastigiate trees may be suitable where a round-headed tree would not be.

(iii) *Space requirements*
Although the area required by a tree will largely depend on its size and shape, the question of root space must also be taken into account.

(iv) *Deciduous and evergreen trees*
The choice between these trees will depend largely on their characteristics.
Deciduous trees:
Drop their leaves in winter.
Are lighter in appearance than evergreens and in winter do not obstruct light to any great extent.
Some provide attractive flowers in the spring while others produce striking colours in the autumn.
Evergreens:
Drop their leaves or needles over a longer period than deciduous trees.
Are darker and cast more shade than deciduous trees.
Although more sombre in appearance can provide an excellent background to flowering or lighter-foliaged deciduous species.

(v) *Special characteristics*

Some trees have special characteristics other than those referred to above which may affect their selection. For example, the roots of poplars can cause considerable damage to buildings and drains if planted near them.

(c) *Sites for trees in urban areas*

There are many sites in towns and their environments where trees can be planted, and these include the following:

Streets and roadsides
New housing estates
Parks
Cemeteries and crematoria
Golf courses
Reservoirs
Industrial sites and factory areas
Industrial workings and waste:
   e.g. Colliery heaps and tips
       Quarries
       Gravel pits
       China clay workings
Special sites:
   e.g. Disused railway tracks and embankments
       Drained canal beds.

3 Trees and tree growth

(a) *Botanical classification*

 (i) The vegetable kingdom has been divided into five divisions and trees have been placed in the division known as Spermatophyta.

(ii) The Spermatophytes are again split into two sub-divisions
   (1) Gymnosperms
   (2) Angiosperms

(iii) The Gymnosperms are chiefly composed of conifers, i.e. pines, spruces, and so on, while the Angiosperms comprise the Monocotyledons, which include grasses, and the Dicotyledons, which consist of the broadleaved trees such as oak, elm and beech.

(b) *The parts of a tree*

 (i) A tree can be divided into four main parts: the roots, the trunk or main stem, the branches (often referred to collectively as the crown) and the leaves.

(ii) The roots have two main functions: to provide the tree with food and water and to act as a support and anchor so as to prevent it from blowing down.

(iii) The trunk or main stem of the tree joins the roots to the crown and raises it above the ground.

(*c*) *How a tree grows*

(i) A tree is composed of a great number of minute cells. These vary in size and type, while some are dead and others alive.

(ii) Young living cells are able to divide and so multiply, and this multiplication occurs chiefly at the ends of the leading shoots and branches and in the cambium layer beneath the bark. By this means the tree is able to increase its height and its girth. Growth also occurs at the ends of the roots.

(iii) Water and mineral food is absorbed by the tree through fine hairs near the apex of the root which are known as root hairs.

(iv) The leaves enable a tree to manufacture starch through the process of photo-synthesis or carbon-fixation, while at the

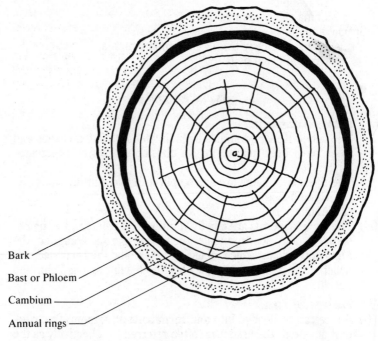

Figure 1.1 Diagrammatic cross-section of trunk of a typical tree

same time the sap conveys food material between the roots and the leaves. Conifers have modified leaves which are popularly known as 'needles'.

(d) Requirements for growth

Certain external conditions are necessary for healthy growth, the chief ones being:
  (i) Suitable temperature.
 (ii) Sufficient supply of water.
(iii) Appropriate supply of food material.
 (iv) Presence of oxygen.
  (v) Although not absolutely essential, light has a beneficial effect on growth.

(e) Assessment of age

 (i) As a tree grows, it produces each year a new layer of wood which is often referred to as the annual ring. In many species, after a tree has been felled, the annual rings can be seen on the stump as a series of concentric circles, and if these are counted the age of the tree can be assessed.
(ii) Certain species of conifers, as for example the pines, produce a ring or whorl of branches on the main stem during each year of growth. By counting the number of whorls the age can be estimated with reasonable accuracy although, as the age increases, the whorls tend to become obscured, especially if the branches fall. Consequently in course of time this method ceases to be of value.

CHAPTER II

The Planting and Care of Amenity Trees

1 *Requisites of a good tree*

(*a*) On the whole trees take a long time to grow and their lives cover a considerable span. It is always worthwhile planting a well-grown tree of the right type; it is never worthwhile planting a poor one.

(*b*) In assessing the merits and suitability it should be kept in mind that a tree ought to be:
(i) Healthy and well grown.
(ii) Well rooted with a good fibrous root system.
(iii) Well proportioned and not drawn up, top heavy, or weak-stemmed.
(iv) Well furnished with branches (if applicable).
(v) Propagated from the best and most healthy stock and true to type and name.
(vi) Transplanted regularly, in the nursery stage.
(vii) Suitable for the situation in which it is to grow
    (1) as regards its ultimate height, crown development and habit;
    (2) in as much as it should enhance its surroundings and not appear out of place.

(*c*) The attention and care which is given to lifting, packing and transporting trees will be reflected in the condition of the trees when they reach their destination.

2 *Types of trees*

Various names are applied to nursery stock in order to indicate the type, size or shape, and some of these are given below.

(*a*) *Seedlings*
(i) These are the smallest plants. They have remained in the seedbed and have not been transplanted, and are usually about 7 cm (3 in.) in height.

(ii) Plants that have been in the seedbed for one year are known as '1-year seedlings' and are referred to as $1 + 0$, while those that have been there for two years are termed '2-year seedlings' or $2 + 0$.

(*b*) *Transplants*
 (i) These are seedlings which have been transplanted, usually two or three times.
 (ii) Those which have been two years in the seedbed and one year in the transplant line are referred to as $2 + 1$.
 (iii) Where these have been grown in the seedbed for two years and in the transplant line for a further two years, they are described as $2 + 2$.
 (iv) Young trees which have been twice transplanted are known as $2 + 1 + 1$, $2 + 2 + 1$, or whatever combination is applicable.
 (v) Transplants are usually from 20 cm (8 in.) to 60 cm (24 in.).
 (vi) Where seedlings have had their roots undercut instead of being transplanted, this is indicated by the letters 'u/c' or by (for example) 2 u 1.

(*c*) *Whips*
 (i) These are so called on account of their single leaders, which are considered to resemble a whip.
 (ii) They are sold according to their size, of which the following are typical:
  60–90 cm (2–3 ft)
  90–120 cm (3–4 ft).

(*d*) *Feathered trees*
 (i) These are trees on which the side branches or 'feathers' have been retained, and this makes a tree more attractive when planted as an individual specimen.
 (ii) The sizes vary but they are usually about 1.2 m (4 ft) to 1.8 m (6 ft) in height.

(*e*) *Standards*
 (i) Standards are trees, the boles or stems of which are clear of side branches.
 (ii) A number of descriptive names are used by nurseries to indicate the size, robustness or type of tree, and these include the following:

Three-quarter standards	Heavy standards
Half standards	Extra heavy standards
Short standards	Semi-standards
Light standards	Weeping-standards.

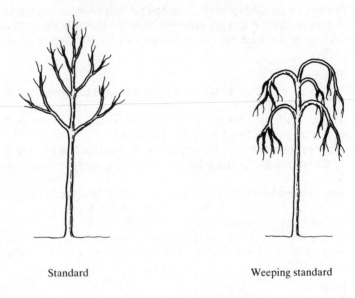

Standard

Weeping standard

Bush

Feathered

Figure 2.1 Types of trees

(iii) As to size, a standard has a clear bole or stem for a minimum of 1.8 m (6 ft) from ground level to the lowest branch.

(*f*) *Large trees*
These can be broadly divided into the following:

  (i) *Advanced nursery stock trees*
  After growing for three to four years in the nursery, they are again transplanted and grown on for a further three or four years.

 (ii) *Semi-mature trees*
  These are grown until they have reached a height of 10–12 m (30–40 ft). On account of their large size, transplanting can be a difficult operation which is not always successful.

(iii) *Instant trees*
  This is a popular name which is applied to the larger-size trees which have been specially grown for transplanting.

(*g*) *Conifers*
Conifers should be well-furnished, that is to say that the branches and foliage should extend for the whole height of the tree, as appropriate to the species concerned.

Note. Full details relating to nursery material will be found in British Standard 3936, 1980 – *Specification for Nursery Stock*: Part 1, *Trees and Shrubs*.

3 *Sources of supply*

Trees may be obtained in the following ways:
  (*a*) By raising in a nursery from seed, cuttings or grafting.
  (*b*) By purchasing seedlings or small plants and growing them on in the nursery until they have reached the desired size.
  (*c*) By purchasing trees of the required size and planting them in their final position.
  (*d*) In the case of poplars, either by purchasing cuttings which are planted in a nursery or by purchasing rooted setts. These can either be planted out or placed in the nursery for a time if larger plants are required.

4 Season for planting

(*a*) Trees should be planted during the winter, when growth has ceased and after the trees have 'hardened off'. In a normal year this can be assumed to be from November to March (inclusive). It is, however, possible to start a little earlier and continue into April if weather conditions permit.

(*b*) Ideally planting should be carried out when the soil is still warm, but this is not always possible.

(*c*) Planting should not be undertaken:

(i) When the ground is excessively wet or waterlogged.

(ii) In cold frosty weather.

(iii) In a cold drying wind.

(*d*) In the case of pot-grown or container trees, planting can be done out of the normal season and consequently the planting period can be extended very considerably. However, if planted during the growing season, they should be watered regularly.

5 Soils

(*a*) *Soil types*

(i) Of all the factors which affect the successful establishment and subsequent growth of a tree, soil is probably the most consistently important.

(ii) Broadly speaking, soils can be divided into five types:

        Clay   Loam

        Sand   Chalk and limestone.

        Peat

(iii) *Clay soils*

(1) Consist of very small particles.

(2) Readily retain moisture, are heavy, cold, sticky and very difficult to work when either very wet or very dry.

(3) Are greatly affected by moisture and expand when wet and shrink when dry, often producing cracks in the soil surface.

(4) Are potentially fertile and are always improved by drainage.

(iv) *Sandy soils*

(1) Can be regarded as the opposite of clay soils and consist of relatively large soil particles.

(2) Are light and easy to work in any season.

(3) Do not retain moisture and in dry weather tend to dry out and 'burn'.

(4) Are not naturally very fertile and require 'feeding', since they are often deficient in potash and nitrogen. They are improved by the introduction of organic matter.

(v) *Loams*

(1) Are a mixture of clay and sand and are described as sandy loam, medium loam, heavy loam, according to the proportion.

(2) Combine the best features of clays and sands and normally produce very fertile soils which are not difficult to work.

(3) Provide suitable soils for almost all species of trees.

(vi) *Chalks and limestones*

(1) Contain a high percentage of calcium carbonate.

(2) Are reasonably easy to work when dry, but chalk soils become sticky when wet.

(3) Require organic manures and potash to increase their fertility.

(4) Are not acceptable to trees which are calcifuges or 'lime haters'.

(5) Can produce chlorosis through lime-induced deficiencies. For further information on this matter reference should be made to *Pathology of Trees and Shrubs* by T. R. Peace (Oxford University Press, 1962) and to *Diseases of Forest and Ornamental Trees* by D. H. Phillips and D. A. Burdekin (The Macmillan Press Ltd, 1982).

(vii) *Peat*

(1) Contains a high proportion of organic matter and is very acid.

(2) Usually requires extensive drainage, and when this has been done can become very fertile, especially when phosphate is applied.

(*b*) *The acid–alkaline association*

(i) Soils vary in their degree of acidity or alkalinity and this is assessed by their pH value (potential of Hydrogen).

(ii) A scale of these values has been drawn up extending from 1 to 14, with 7 representing neutral conditions. All readings less than 7 are acid and all above are alkaline. Very acid conditions are indicated by a value of about 4·5, moderately acid by 5·5 and slightly acid by 6·5.

(iii) The pH of a soil sample can be found either by using a soil pH meter or by means of a soil testing kit. These can usually be obtained through any large garden centre or horticultural store.

6 *Preparation for planting*

(*a*) The amount and extent of the preparatory work prior to planting depends on the size of the planting operation. This could consist of planting:
  (i) Small individual trees, which might be standards, half standards, feathered trees and so on;
 (ii) Large trees of a height of up to 9 m (30 ft);
(iii) Semi-mature trees of over 9 m (30 ft);
 (iv) Limited groups of small individual trees;
  (v) Shelter belts or other blocks of trees of the size normally used in forestry, e.g. 20 cm–46 cm (8 in.–18 in.).

(*b*) In the case of the small trees, whether planted individually or in groups, the following action should be taken:
  (i) All surface growth such as brambles should be removed and the ground broken up and thoroughly dug over.
 (ii) If soil conditions are not satisfactory, humus or selected top soil should be worked into the ground.

(*c*) Similar action should be taken for the larger trees referred to in (*a*) (ii) and (iii) above, but because of the increased size of the trees, the area of ground covered will be much more extensive. The transplanting of large trees is dealt with in Chapter III.

(*d*) In the case of shelter belts and the planting of forest trees, the preparation will simply consist of removing any competing surface vegetation such as gorse, thorns, brambles and so on, which would otherwise interfere with the young trees.

(*e*) In all cases, irrespective of the size of the trees, it is essential that the planting site is propery drained. Failure to do so can lead to set backs, losses and windblow.

7 *Spacing*

(*a*) When planting individual specimen trees, it is vital to allow sufficient space for their ultimate growth and development. The effects of failing to appreciate this point can be seen in numerous collections of trees at the present time.

(*b*) In order to decide what spacing should be adopted between trees, it is necessary to visualize the size of the tree concerned when it is fully grown. Assistance in this, can be obtained from *Tree Form, Size and Colour* by J. St Bodfan Gruffydd (1987).

(*c*) However, this means that in the early years the site may appear bare and open. In order to diminish this effect two courses are open:
(i) To plant next to each other different species of trees which will attain varying heights when mature.
(ii) To plant more trees than are necessary, and during the course of years to remove a substantial proportion of them, so as to allow the remainder to develop. The risk which attends this method is that the excess number of trees may not in fact be removed when the time comes.

(*d*) Where shelter belts or small blocks of trees are to be established, trees should be planted as in forestry, when the spacing will be between 1.5 m (5 ft) and 3 m (9½ ft), the wider spacing being adopted for faster-growing species.

8 *Planting*

(*a*) *Preliminaries to planting*
(i) The method of packing trees depends on their size. Large individual trees are generally packed in moss and straw, while trees for forestry planting are usually packed in plastic bags.
(ii) On arrival the trees should be unpacked and heeled or sheughed in, until they are to be planted. Small trees can be left in their plastic bags for a short time, but all packed trees run the risk of heating, and this must be avoided at all costs. To save time, trenches can be dug before the trees arrive.
(iii) When heeling in small trees, more particularly conifers, care must be taken to spread them out as much as possible so as to minimize the risk of heating. If the trees show any sign of dryness when they are unpacked, they should be dipped in water and the supplier informed immediately.
(iv) However, in the case of trees which have been lifted with a ball of soil around their roots, care should be taken to avoid the ball being disturbed.
(v) If the trees arrive when the ground is frozen, they should be placed under cover in a shed, and straw, bracken or other similar material placed over their roots.

(b) The planting hole

(i) Assuming that any preparation which may be necessary has been done, the first action is to dig a suitable hole or pit to receive the roots of the tree.

(ii) The width of the hole will depend on the size of the tree and the extent of its roots. The pit should be sufficiently large for the roots to be amply spread out.

(iii) The depth of the hole will again depend on the root system, bearing in mind that when the operation has been completed, the root collar of the tree should be level with the surface of the ground.

(iv) This means that a standard tree will probably need a hole about 107 cm (3 ft 6 in.) in width and from 46 cm to 76 cm (1 ft 6 in. to 2 ft 6 in.) in depth. Larger trees will need larger holes.

(v) Small trees for forest planting are either notched in on the lighter soils or pit planted.

(vi) When planting on an embankment or other sloping ground, a 'shelf' should be cut out of the slope.

(c) Planting

(i) After the planting hole has been dug, the bottom of the pit should be forked over and thoroughly loosened, and where the soil is poor or very heavy, a mixture of loam and leaf mould should be introduced. The more care and trouble which is taken in planting the tree, the more likely is it to flourish.

(ii) The tree should then be positioned in the pit and if it is to be staked, the stake should be driven at this stage.

(iii) Soil is then spread around the roots and if necessary worked round them by hand. In the case of the larger trees the soil must be firmed down continually as it is back filled. This is continued until the operation is complete and the soil level with the root collar of the tree.

9 Staking and tying

(a) General

(i) The method which is adopted in staking and tying a tree depends to a large extent on its size and type; a densely foliaged conifer can present a different problem to a standard broadleaved tree.

(ii) Although staking is frequently necessary, it should be carried out as inconspicuously as possible, compatible with its effectiveness.

(iii) Trees which have been staked must be inspected at regular intervals so as to ensure that the stakes are still firm, have not broken off, and that the ties are not too tight or too loose so as to cause chafing.

(iv) Further information on staking trees will be found in Forestry Commission Arboriculture Research Note 40/87/ARB, *Tree Staking*.

(*b*) *Objects in staking*

The objects in staking a tree are as follows:

(i) To prevent the tree from rocking in the wind and thereby loosening its root hold.

(ii) In more boisterous conditions, to prevent the tree being blown down or the main stem being broken.

(iii) To provide some resistance to petty vandalism and damage.

(iv) To prevent distortion of the tree through lack of support.

(*c*) *Staking*

(i) Stakes should not be larger than is necessary. If they are, they not only appear clumsy but they are also more expensive.

(ii) Stakes should always be impregnated under pressure with a preservative such as Celcure. Timber which is treated with this type of preservative is clean and easy to handle and does not bleed or weep. If treated in this way, stakes will last for many years and can be used over and over again.

(iii) Staking of trees may be carried out by

               Single stakes   Prop stakes
               Double stakes   Guying.

These methods are described below.

(iv) *Single staking*

(1) Single stakes are suitable only for comparatively small trees or those which do not have a large crown and so offer little resistance to the wind (see Figure 2.2).

(2) The stake should always be placed in position at the time of planting, after the tree has been positioned in the planting hole and before the hole is filled in. It should be on the windward side of the tree to reduce the risk of the tree being blown against the stake.

(v) *Double staking*

(1) Double staking provides a more rigid means of supporting a tree, and in the case of large trees it is to be preferred to single staking. However, it is more expensive (see Figure 2.3).

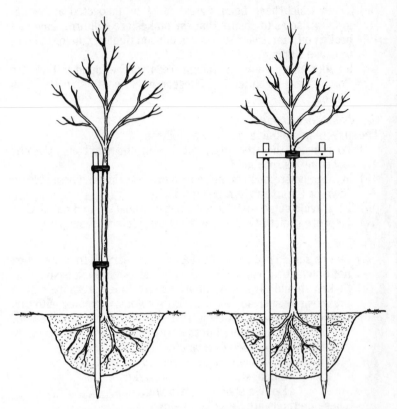

Figure 2.2 Single staking   Figure 2.3 Double staking

(2) Two stakes are driven in about 30–46 cm (1 ft–1 ft 6 in.)
    apart and a cross-member fixed between the two stakes at
    the top of the stakes. One or more additional cross-
    members can be fixed at suitable distances between the
    top member and the ground level. The stem of the tree is
    then secured to the cross-member by ties.
(3) An alternative method is to fix the cross-members in
    parallel pairs, one on either side of the stem. However,
    care must be taken to ensure that there is adequate space
    for the stem so that it is not constricted.
(4) When double staking is adopted the cross-section of the
    stakes used can be slightly smaller than when a single
    stake is used.

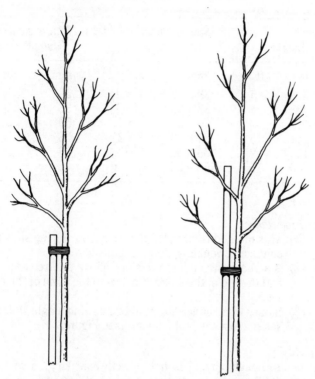

Figure 2.4 Correct method   Figure 2.5 Incorrect method

(5) The size of single stakes will depend on the size of the tree which they have to support. When staking standard trees care should be taken to prevent the top of the stake projecting into the crown of the tree or damaging the side branches (see Figures 2.4 and 2.5).

(6) The distance which a stake should be driven into the ground depends on the total length of the stake, which in turn depends on the height of the tree. A stake which is 1.5 m (5 ft) long should be driven into the ground for about 46 cm (1 ft 6 in.); a 2.4 m (8 ft) stake would have about 76 cm (2 ft 6 in.) in the ground.

(7) The following is a guide to the length and thickness of stakes:

Length		Square stakes (cross section)		Round stakes (diameter)	
m	ft	mm	in.	mm	in.
1.5	5	38 × 38	$1\frac{1}{2} \times 1\frac{1}{2}$	38	$1\frac{1}{2}$
1.8	6	38 × 38	$1\frac{1}{2} \times 1\frac{1}{2}$	38	$1\frac{1}{2}$
2.1	7	50 × 50	2×2	50	2
2.4	8	50 × 50	2×2	63	$2\frac{1}{2}$
2.7	9	63 × 63	$2\frac{1}{2} \times 2\frac{1}{2}$	63	$2\frac{1}{2}$
3.0	10	63 × 63	$2\frac{1}{2} \times 2\frac{1}{2}$	76	3

(vi) *Prop staking*
   (1) This method is intended to be used on sloping sites where single and double staking is difficult.
   (2) A stake is driven into the slope above the tree at an angle and the top of the stake attached to the stem of the tree by a tie.
   (3) Stakes of a similar size to those used in single staking are satisfactory for prop staking (see Figure 2.6).

(vii) *Guying*
   (1) Although guying is not, strictly speaking, a method of staking since no stakes are used, it has been included in this section partly as a matter of convenience and also because it is a recognized method of securing trees (see Figure 2.7).
   (2) Guying is normally used for stabilizing trees which are too large for staking, for semi-mature trees or for those which, for one reason or another, are difficult to stake.
   (3) Four anchoring points which can be either of treated timber or angle iron are driven in at the four corners of an imaginary square surrounding the tree. These anchoring points function in a similar way to tent pegs.
   (4) Wire guys are attached indirectly to the stem of the tree and to each anchoring point, and by means of these the tree is held securely in position.

(*d*) *Tying*
   (i) The term 'tying' is used to denote the means by which a tree is attached or fastened to a stake or pair of stakes, or guy wires.

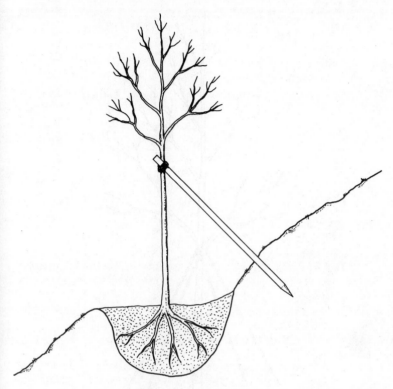

Figure 2.6 Prop staking

(ii) The following are some of the more usual tree ties which are to be seen in use:

(1) *Treated cord*

This is not a satisfactory method of tying, since it lacks strength and tends to cut into or chafe the stem of the tree. Its use should be avoided except in the case of very small trees, prior to being planted out.

(2) *Hessian or canvas strips*

Where a large number of trees are to be staked and tied, as for example in a young plantation, hessian or canvas is a reasonably cheap and adequate method of tying. Canvas is stronger than hessian and neither material causes chafing. Hessian has only a comparatively short life, but canvas will last longer.

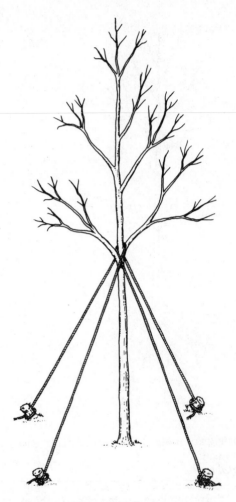

Figure 2.7 Guying

(3) *Foam rubber and string*
    Although more satisfactory than string alone, this method
    is not entirely satisfactory. Care must be taken to avoid the
    rubber pad slipping or the string cutting through the pad.
(4) *Plastic and rubber ties*
    There are now several designs of this kind of tie on the
    market. Broadly speaking they are of two main patterns,

namely those which are not nailed or fastened to the stake and those which are. In both patterns a rubber block is provided between the stem of the tree and the stake to act as a cushion.

(5) *Protected wire*
This is usually effected by passing a guy or tying wire through a short length of old rubber or plastic hose. The object is to prevent the wire cutting into the bark and to reduce abrasion.

(iii) It is advisable to use two ties on standard trees, and where they are large, three ties may be necessary. Additional ties can also be provided where double staking is adopted (see Figures 2.2 and 2.3).

10 *After-treatment*

For the first one or two years after planting special care and attention will be well repaid.

(*a*) *Treading*
 (i) No matter how firmly a tree is planted, during the first year after planting, treading or firming should be carried out.
(ii) A tree can become loose due to several causes, such as wind, frost, activities of moles and dry conditions at the time of planting.

(*b*) *Mulching*
If the spring and summer following planting are dry, a surface mulch of well-decayed leaf mould or other organic matter should be applied to a depth of 8–10 cm (3–4 in.)

(*c*) *Watering*
 (i) In addition to mulching, watering may also be necessary if the first spring and summer are dry. The benefit of watering will be greatly increased if carried out in conjunction with mulching, especially on dry soils.
(ii) When watering, it is important to ensure that the tree receives a sufficient quantity. In dry weather or on dry sites the soil around the tree should be thoroughly soaked. A large standard will need 27–30 litres (6–7 gallons) in such circumstances.
(iii) In dry weather evergreens can be helped by spraying the foliage, but this should not be done in hot sunshine or when there is risk of frost.

(*d*) *Weeding*

(i) After planting, an area (about 1 m or 1 yd. in diameter) immediately surrounding the base of the tree, should be kept free of grass and other weeds.

(ii) This can be done by:
  (1) cultivating the area by hoeing, or
  (2) applying a herbicide, or
  (3) laying a sheet of black polythene or other suitable material.

(iii) Full details will be found in Forestry Commission Handbook No. 2 – *Trees and Weeds*.

(*e*) *Pruning*

(i) Recently planted trees should be gently pruned when the spring growth begins.

(ii) Any long shoots can be shortened and evergreens should have their leaf surfaces reduced.

(iii) The object of this work is to encourage compact healthy growth and to ensure a proper balance between the foliage and the roots.

(iv) The wider aspects of pruning, and more especially the removal of larger branches, are dealt with in Chapter VI.

11 *Protection*

Injuries, and the steps to be taken to protect trees from them, are considered in Chapter IV.

CHAPTER III

Transplanting Large Trees

1 *Introduction*

(*a*) *Present-day needs*

(i) The accelerated tempo of this century is reflected in numerous ways, and not the least of these is in the increasing demand for immediate results. While these can often be achieved with man-made products, nature does not readily acquiesce. However, through the medium of transplanting large trees, it is possible, to some extent, to take time by the forelock.

(ii) The increasing demand for 'ready-made' trees is due to several factors including the following:

(1) The rapid extension and development of urban areas which are completed long before a tree could grow to a substantial size.

(2) The need to soften the hard lines of modern architecture, more especially the 'glass and concrete box' type of building.

(3) The fact that increasing attention is being paid to amenity, especially in urban areas.

(*b*) *Designation of trees for transplanting*

Large trees which are used in transplanting have been referred to by various names, of which the following may be mentioned.

(i) *Mature trees*

This term should not be used, as it is inaccurate and misleading. Trees which have reached full maturity would be unsuitable for transplanting and are not used.

(ii) *Semi-mature trees*

These are trees that have reached such a size that it is generally necessary to provide a root-ball, so as to ensure satisfactory results when transplanting them. They are normally between 7 m (23 ft) and 16 m (52 ft) in height and from 0.25 tonnes to 10.16 tonnes in weight.

(iii) *Instant trees*
This is the popular name for semi-mature trees, which reflects the modern aspirations for quick results.

(c) *Historical background*

(i) Although the transplanting of large trees has received considerable attention during the past twenty years, there is in fact nothing new or original in the idea. John Evelyn refers to the subject in detail in *Sylva* (2nd edn, 1670) and mentions the planting in Devon of 'oaks as big as twelve oxen could draw to supply some defect in an avenue'.

(ii) In 1827 Sir Henry Steuart wrote *The Planter's Guide*, in which he describes at very great length the methods which he adopted to move large trees on his estate at Allanton in Scotland. His two-wheeled transplanting machine was very similar in appearance to a timber bob or janker and was probably developed from it.

(iii) Some years later, in 1844, Colonel George Greenwood was the author of another book, entitled *The Tree Lifter*, which gave a full account of a machine which he had designed for the transplanting of large trees. The design of his machine, which was also two-wheeled, differed from Steuart's in that the tree was transported in an upright or vertical position, whereas in Steuart's machine the tree travelled in a horizontal position, i.e. with the trunk parallel to the ground.

(iv) William Barron, who was head gardener at Elvaston Castle, wrote a small book in 1852 under the title of *The British Winter Garden*, which he described as 'a practical treatise on evergreens . . . and their mode of propagating, planting and removal from one to fifty feet in height'. Barron used two machines, a four-wheeled one for moving large trees and a three-wheeled one for small trees. In both machines the tree travelled in a vertical position. Barron subsequently founded a firm which for many years specialized in transplanting large trees.

2 *Trees for transplanting*

In deciding whether a tree is suitable for transplanting to a given position, the following points should be considered:

(a) *Purpose of the operation*
In the first place it must be decided what is the purpose for which the tree is being moved. It may be in order to create a feature, to fill

a gap, to provide a screen, to add colour or to constitute a background. The tree may have to stand by itself or it may be one of several which together will produce the intended effect.

(b) The planting site
Much depends on the site on which it is intended to plant the tree. As in the more usual forms of planting the questions of site, soil, aspect, elevation, exposure, growing space and so on must be taken into consideration.

(c) The condition of the tree to be moved
The state of the tree before transplantation must be borne in mind. Its shape, height, crown development, rooting system, age, health and general condition must all be considered before finally deciding to move it.

(d) The species of tree
  (i) Some species of trees are more suitable for transplanting than others. Broadly speaking, the following are most adaptable: alder, elm, horse chestnut, lime, maple, plane and poplar.
 (ii) The less adaptable species include ash, beech, birch, holly, hornbeam, Lawson cypress, the silver firs, spruce, sweet chestnut, walnut, willow and yew.
(iii) The least adaptable include the true cedars, larch, magnolias, oak and the pines.

3 Preparation for transplanting

(a) Time of year for transplanting
  (i) Transplanting should be carried out during the winter between November and March, after the previous season's growth has stopped and before the next season's growth has commenced.
 (ii) Transplanting should not be carried out in dry cold weather, during frosty conditions or at any time when the weather would be unsuitable for normal planting operations.

(b) Root preparation
  (i) Trees should receive preparatory treatment some time before they are moved.
 (ii) This treatment consists chiefly in the reduction of the root system in order to encourage the development of fibrous roots and also to facilitate the transportation of the tree.
(iii) Those trees which are the easiest to move require a minimum interval of one growing season between the time when the

preparatory work has been carried out and the actual transplantation. The remainder should have two growing seasons between preparation and moving, but it must be emphasized that the longer the period of preparation the more satisfactory are the results likely to be. This allows the tree time to form new fibrous roots which are essential for its survival and well-being.

(iv) Preparation consists of digging a trench around the tree so as to cut the roots which grow outwards from it and roughly parallel to the surface of the surrounding ground. This trench marks the limit of the ultimate root-ball which will contain the roots of the tree when it is lifted.

(v) On heavy soils and in the case of species which are able to adjust their root systems less readily, the encircling trench should not be completed in one operation. The circumference of the trench should be divided into four to eight segments and only the alternate segments are cut out at one time. An interval of one year is allowed to elapse before the second set of segments is removed (see Figures 3.1, 3.2 and 3.3).

(vi) After the trench has been dug it may be treated in one of the following ways:
(1) Filled with loam and humus.
(2) Filled with cinders.
(3) Unfilled and left open.
In the first method, new root growth will develop in the newly filled trench, while in the other two methods root development should occur within the undisturbed earth of the root-ball. This means that the young root growth is less liable to damage during transplanting. However, the presence of the loam will encourage the growth of new roots and root hairs which will not be formed so readily in the existing soil surrounding the roots. Steps should be taken to ensure that the filled loam is firm and if necessary watered.

(vii) In the case of any roots which are thicker than 2 cm ($\frac{3}{4}$in.) diameter being severed, they can be trimmed so as to remove any ragged or shattered surfaces. However, in practice it is not always possible to take such action.

(viii) It is important that as little movement of the tree as possible should take place after the preparatory trench has been dug, and if necessary resort must be had to guying.

(ix) The distance at which the trench is dug from the bole of the tree, and therefore the diameter of the ultimate root-ball, will depend on a number of factors. Amongst these are the type of

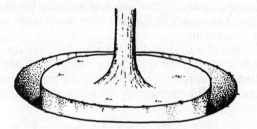

Figure 3.1 Trenching before removal

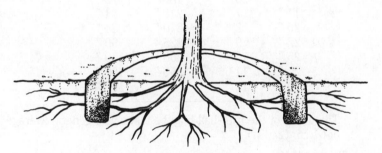

Figure 3.2 Root pruning

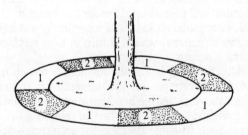

Figure 3.3 Trenching by segments
1. Removed first.   2. Removed one year later

soil, the species of tree, its root system, its age, girth and height. Consequently it is not wise to lay down any hard and fast dimensions.

(x) However, the following relationship between the diameter of the bole of the tree (at a point where it is clear of the root buttresses: say, 45 cm (18 in.) above ground level) and the diameter of the preparatory trench or root-ball may be taken as an approximate guide:

Diameter of bole at 45 cm (18 in.)		*Diameter of trench or root ball*	
10 cm	4 in.	120 cm	48 in.
12 "	5 "	150 "	60 "
15 "	6 "	180 "	72 "
17 "	7 "	210 "	84 "
20 "	8 "	210 "	84 "
22 "	9 "	225 "	90 "
25 "	10 "	250 "	100 "
27 "	11 "	250 "	100 "
30 "	12 "	270 "	108 "

The depth of the root-ball may be taken as a general average of 61 cm (24 in.).

(c) *The root-ball*
  (i) Immediately prior to lifting, the roots of a prepared tree are encased in a mass of soil. The solidity of this mass will largely depend on the type of soil concerned. A heavy soil will hold together to a far greater extent than a sandy one.
 (ii) Apart from the nature of the soil, a large root-ball of earth would never remain intact throughout the process of loading, transporting and unloading. In order to overcome this difficulty the ball of earth and roots is held together by a skirt or surround of hessian or canvas, which in turn is held in place and given additional strength by ropes or chains.
(iii) To enable the earth-ball to be formed, a trench must be dug beyond the perimeter of the preparatory trench which is described above under the heading *Root preparation*. The object of this trench is not only to assist in releasing the tree, but at the same time to avoid damage to, or interference with, the new roots which have been formed in the medium of the preparatory trench.

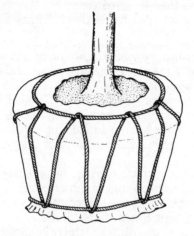

Figure 3.4 Root-ball held by ropes

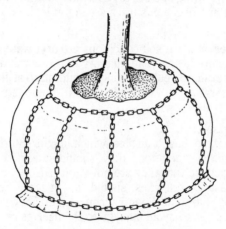

Figure 3.5 Chains holding root-ball

(iv) After excavation, a hessian or canvas skirt referred to above is placed in position, and when this is complete the roots beneath the ball are severed. This can be done in two ways:
   (1) By pulling the tree over at an angle so that the underlying roots are broken off.
   (2) By passing a winch rope under the tree so that it severs the roots, its action being similar to that of a cheese wire cutting through cheese.

The tree can then be lifted clear, but this work is described later in this chapter under Section 5, Lifting and transporting.

(*d*) *Anti-transpirants*
  (i) In order to reduce excessive transpiration and wilting caused by the severing of the roots, the undersides of the leaves may be sprayed with special liquids.
 (ii) There are, however, difficulties in this kind of treatment. A spray of this sort is not easy to apply successfully to a semi-mature tree. Over a large surface the application may be too light, or too heavy, while some parts may escape treatment entirely.
(iii) This form of preparatory treatment originated in the United States, where transplanting is carried out in summer as well as winter. Also known as anti-desiccant sprays, there are several available including Synchemicals S.600.
(iv) The problems of spraying can largely be avoided by restricting the transplanting of trees to the winter months, when growth is dormant and deciduous trees are not in leaf.

(*e*) *Watering*
In dry weather or when soil conditions are dry, watering should be carried out before lifting. The quantity of water will depend on the conditions prevailing, but watering should start at least 48 hours before lifting is due to take place and should be continued at requisite intervals.

(*f*) *Branch restriction*
  (i) When a tree has branches which might interfere with the transplanting operation, they should either be removed or lashed back so that they are not damaged.
 (ii) If necessary, branches should be encased in hessian by wrapping it around the limbs concerned. They should then be secured so as to avoid damage to themselves or to other parts of the tree.
(iii) In addition, if a tree has a large crown, this should be reduced before transplanting so as to reduce the demands of the branch and leaf system on the roots.

4 *Preparation of the planting site*

(*a*) *Drainage*
  (i) The planting site must be well drained, and steps should be taken to achieve this.

(ii) Some species, such as alder and poplar, are prepared to accept moister conditions than others but this does not mean that they should be planted on wet, poorly drained sites. Stagnant soil water is fatal for tree survival.

(iii) Where the water table is high and the removal of excess surface water is difficult, resort may be had to planting on mounds or heaps of soil raised above the surrounding land.

(b) *Planting pits*

(i) The hole or pit in which a tree is to be planted should be carefully prepared. The bottom should be well dug over, broken up and then firmed down.

(ii) The hole must be large enough to allow for selected top soil or loam to be placed between the root-ball and the bottom and sides of the pit. In order to achieve this, the pit should have a depth and radius approximately 30 cm (12 in.) greater than the ball.

5 *Lifting and transporting*

(a) *Lifting and loading*

(i) Having carried out the necessary preparation for transplanting, as described in Section 3 of this chapter, the tree should be ready for lifting.

(ii) This can be done in two ways:
By hand
By machine.

(iii) *By hand*
(1) Although this method is described as being by hand, the actual raising of the tree is carried out by mechanical means such as a crane or hoist.
(2) After the root-ball has been formed, hoisting ropes are passed around the base of the stem of the tree, which is the point of balance. Care must be taken to ensure that the rope does not cut into the stem, and the necessary precautions must be taken to prevent this.
(3) The actual loading and unloading may be carried out by:
a. Independent crane which loads the tree on to a carrying vehicle.
b. Hoist or crane which is mounted on the same vehicle which is to transport the tree.
c. Specially designed tree-moving trailer which is self-loading. The principle adopted in loading is to back the

trailer to the tree and then up-end it. The tree is made fast, and the trailer is then brought back to horizontal position, bringing the tree with it.

(iv) *By machine*

With this method the tree is dug up and replanted entirely by machine.

(*b*) *Transporting*

Today trees are normally transported on lorries or by a specially designed trailer as referred to above.

6 *Replanting*

(*a*) The first step in replanting is to ensure that the planting hole has been properly prepared as regards siting, draining, depth, diameter and so on. It is wise to check the dimensions of the root-ball against those of the prepared hole.

(*b*) If the tree has to be planted so that it faces a particular aspect, steps should be taken to ensure this at the time of unloading.

(*c*) The tree should be planted at the same depth at which it was growing before its removal. A slight ridge of soil may be formed around the circumference of the planting hole (after it has been filled in) so as to form a 'dam' to retain water after watering.

(*d*) When filling the cavity between the root-ball and the sides of the planting pit, care must be taken to ensure that the soil is really firm. Filling should be done in layers, each layer of soil being firmed down before the next one is added.

(*e*) Only the best top soil should be used for filling, and this may be mixed with a proportion of well rotted leaf mould or peat if necessary.

7 *After-treatment*

(*a*) *Watering*

 (i) In a dry season or if the soil is such that it remains dry under normal weather conditions, watering should be undertaken.

(ii) A newly transplanted tree will also benefit if the foliage is sprayed during the summer.

(*b*) *Mulching*

 (i) Mulching will assist in maintaining moist conditions over the root area.

(ii) The material used for this purpose may comprise leaf mould, bracken, peat, grass cuttings, cut weeds and so on.

(iii) The amount of material used should be such as to produce a layer of not less than three inches.

(c) *Pruning*

(i) In the course of moving a tree a certain amount of damage is inevitable.

(ii) After transplanting, any broken branches should be removed and the cut surfaces treated as described on page 80, *Treatment of wounds*.

(iii) Any damage which has occurred to the bark should be pared away.

(d) *Wrapping*

(i) To reduce the risk of excessive drying and the effect of hot sun, the main stem and larger lower branches of a newly transplanted tree should be adequately protected.

(ii) This may be done by wrapping the appropriate parts of the tree in suitable material. Such material includes light canvas or hessian strips, straw or straw ropes, or specially prepared proprietary wrappings.

(e) *Stabilizing*

(i) After the replanting operation has been completed, a transplanted tree has very little resistance to the wind. This is due to the fact that the roots have been reduced to the limits of the root-ball which on account of its size and shape provides little if any stability.

(ii) In order to overcome this problem it is necessary to provide some means of maintaining the tree in a fixed position so that it does not rock in the wind. This may be done in the following ways:

Staking

Ground anchors

Guy wires.

(iii) *Staking*

This is suitable only for small trees and is not really applicable to semi-mature trees. The subject is dealt with in Chapter II.

(iv) *Ground anchors*

(1) The principle which is adopted when ground anchors are used, is to hold the root-ball in position and so prevent movement. On the other hand where guy wires are used,

the purpose is to hold the bole or trunk in position and so prevent its acting as a fulcrum on the root-ball.

(2) There are two main types of anchors:
   a. Stakes driven into the firm ground around the root-ball.
   b. Anchors sunk below the surface of the ground.

(3) In each case wires are passed between the stakes or ground anchors so that they traverse the root-ball and so hold it in position.

(4) In order to prevent these wires cutting into the ball, a wooden frame or wooden bearers should be placed on top of the root-ball and the wires led over these.

(5) All the timber which is used for stakes, bearers or ground anchors should be treated under pressure with a preservative such as Celcure.

(6) Figures 3.6–3.9 show the details of the above methods.

(v) *Guy wires*

(1) The use of guy wires or 'guying' will give greater stability in the case of large trees than ground anchors.

(2) The procedure adopted is to rig guy wires from a point on the main stem to three or four anchor stakes which have been driven into the ground. The point at which the wires are fixed to the bole should be from one half to two thirds of the height of the tree above ground level.

(3) If there is a conveniently placed fork or group of branches growing out from the main stem at the requisite height above ground level, the guy wires may be passed around these. In order to avoid damage to the stem, the end of the wire should be passed through a length of old rubber hose and a pad of foam rubber placed between the hose-covered wire and the stem. To facilitate fixing, this pad can be tied with string, which is then removed after the guys are in position since their tension will hold the pad in place.

(4) If there are no convenient branches, provision must be made for attaching the end of the guys to the stem itself. To do this, a piece of foam rubber not less than 1 inch thick and 6 inches wide should be wrapped around the stem at the required height. This pad can be temporarily held in position with string. A piece of canvas is then wrapped around the foam rubber and finally four or five wooden slats not less than 2.5 cm (1 in.) wide, 18 cm (7 in.) long and 6 mm ($\frac{1}{4}$ in.) thick are placed on top of the canvas. These slats may be attached to the canvas by a tack driven

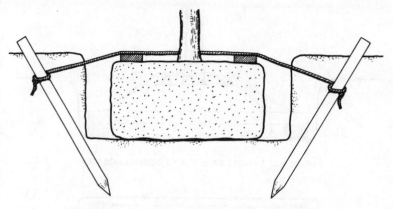

Figure 3.6 Ground anchor with stakes (side view)

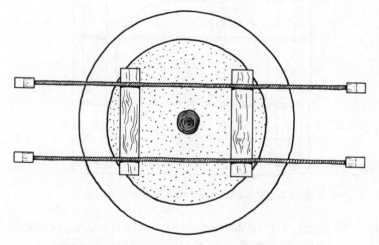

Figure 3.7 Ground anchor with stakes (from above)

through from the inside of the canvas into the slat. The whole is then held in position by two bands of light tying wire.

(5) The guy wires are then fixed around the slats which are held in position by the tension of the guys. The disadvantages of this method are that the slat and canvas pad may slip downwards unless it is wrapped tightly around the

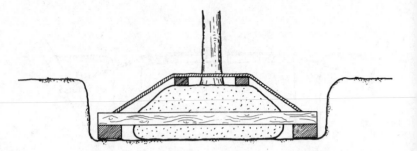

Figure 3.8 Ground anchor with baulks (side view)

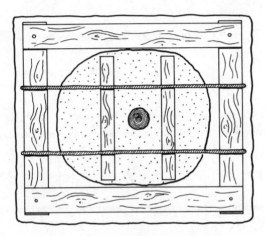

Figure 3.9 Ground anchor with baulks (from above)

stem. If this is done, it will have to be readjusted during the growing season to avoid constriction of the stem.

(6) The stakes should be driven into the ground at a distance from the tree which will give an angle of between 45° and 60° with the surface of the ground. An angle of 45° will give greater stability than one of 60°, but the anchor stakes will then be further from the tree than in the case of a 60° angle. In certain circumstances this can be a disadvantage.

(7) If anchor stakes are considered undesirable, ground anchors can be used instead, but more work is involved and the cost is proportionately greater. Ground anchors are dealt with in section (iv) above.

(8) The tension of the guys can be maintained by means of turn-buckles, by using two wires, inserting a piece of metal or wood between the wires, and operating as in a tourniquet, or simply by driving the stakes further into the ground.

8 *Container-grown trees*

(*a*) These are trees which have been grown in portable containers for the greater part of their lives.

(*b*) When raising trees of this kind, the following points should be borne in mind:

(i) The container must be large enough to ensure a reasonable amount of space for the roots. However, it may be necessary to prune the roots when the tree is first placed in the container.

(ii) Owing to the restricted growing space, frequent watering is essential, but at the same time drainage facilities must be provided to avoid waterlogging.

(*c*) The advantages of container-grown trees are:

(i) The trees can be planted throughout the year and not only during the normal planting season.

(ii) A greater degree of success can be expected especially in the case of those species which are difficult to transplant.

(*d*) However, trees that have been growing in containers for some time so that they have become pot-bound are unlikely to be a success.

9 *Advanced nursery stock trees*

(*a*) Advanced nursery stock trees are those which lie between the limits of the usual nursery grown trees on the one hand and semi-mature trees on the other.

(*b*) Their height ranges from approximately 3 to 6 m (10 to 20 ft) and they are transplanted more frequently than normal nursery stock, with the resulting increase in root development.

(*c*) The advantages claimed for this type of tree are:

(i) Since they produce thicker and therefore stronger stems or boles, they are more able to withstand vandalism and wanton damage.

(ii) Being larger than ordinary nursery stock, they produce a more immediate effect.

(iii) On account of the more frequent transplantings which these trees undergo, the root system becomes more fibrous than ordinary nursery stock. Consequently they are more successful than normal nursery stock.

CHAPTER IV

Tree Injuries and Protection

A *Tree Injuries*

1 *Climatic factors*

(a) Frost

  (i) In Britain the most serious damage is caused by frosts which
      occur in late spring or early autumn. In some cases, however,
      damage may be the result of intense cold in winter, but this is
      not very common.
 (ii) Injury is often due to a quick thaw, and this tends to occur on
      sites with a south or south-east aspect which receive early
      morning sunshine.
(iii) Frost crack is the result of the cambium being injured, and
      although the wound heals over, the timber will generally be
      blemished. Frost rib occurs where repeated frosts cause a
      crack to open up in successive years. Trees with thin bark are
      the most susceptible to this form of damage.

(b) Lightning

  (i) Damage by lightning can be divided into two types: firstly
      scarring or cracking, and secondly complete shattering of the
      tree.
 (ii) It would appear that some species are more likely to be
      damaged by lightning than others, but there is little informa-
      tion available on the matter. However, there seem to be some
      grounds for the following observations:
      (1) Trees which have a smooth bark such as beech and
          sycamore appear to suffer less than trees which have a
          rough bark such as oak. This may be due to the fact that
          smooth bark offers less resistance.
      (2) Large trees are struck more frequently than small ones,
          but this is only what one might expect since they cover a
          larger area.

(3) Old trees tend to be struck more often than young trees. This may be due to the fact that old trees are often large trees.

(iii) Many species have been recorded as having been struck by lightning, but on the whole it seems that broadleaved trees are more frequently damaged than conifers. The following species have been recorded as having been struck: alder, ash, beech, cherry, sweet chestnut, elm, plane, poplar, sycamore, spruce, Douglas fir, larch, Scots pine and Wellingtonia.

(iv) Additional information is available in Forestry Commission Arboriculture Research Note 68/87/PAT – *Lightning Damage to Trees in Britain*.

(*c*) *Snow*

(i) Specimen trees can be severely damaged through snow breaking the leaders or side branches. Evergreen species generally suffer the worst damage since they are clothed in foliage during the winter. The true cedars, e.g. the Cedar of Lebanon, are particularly susceptible owing to the fact that the snow lodges on their horizontal branches, which consequently break off under its weight.

(ii) In addition to detracting from the appearance of the tree, broken branches provide an entry point for fungal infection unless the wounds are dealt with shortly after the damage occurs.

(*d*) *Sun*

(i) Hot sun may cause damage to trees under certain conditions.

(ii) A drying wind and sunshine in late March and April can cause considerable damage to young conifers, especially Lawson cypress, Western red cedar and other trees with this type of foliage.

(iii) The bark of a tree may be injured if it is suddenly exposed to hot sun by the removal of accustomed shade.

(iv) The boles of street trees can also be affected by reflected heat from roads or buildings.

(v) In nurseries young seedlings may be killed in very hot weather unless they are afforded protection.

(*e*) *Water*

(i) Although water can only indirectly be regarded as a climatic factor, it is included in this section for convenience.

(ii) An excess of water:

(1) May be caused by poor drainage or a rise in the water table.

   (2) Can cause death of young trees and produce 'stagheaded-
       ness' in old trees.
(iii) A lack of water:
   (1) May be caused by excessive drainage, a fall in the water
       table or a prolonged drought.
   (2) Can cause death of young trees, die-back of the leading
       shoots and also produce 'stagheadedness' in old trees.
(iv) Further details will be found in Forestry Commission Arbori-
     cultural Leaflet No. 6 – *Trees and Water*.

(*f*) *Wind*
  (i) Wind can cause the following damage:
     (1) Blowing down individual trees.
     (2) Damage to the crown or breaking of the main stem.
     (3) Distortion of the crown especially near the sea coast or in
         exposed situations.
     (4) Excessive transpiration resulting in stunted growth.
 (ii) Although the prevailing wind over much of this country is
      considered to be south-westerly, it must not be forgotten that
      strong winds can blow from any quarter. It is often a wind
      which blows from an unusual direction which causes the
      heaviest damage.
(iii) Trees which have always grown singly generally suffer less
      wind damage than one which after many years is suddenly
      exposed to strong winds.
(iv) Other things being equal, trees growing on a badly drained site
     are more liable to be blown down than those growing on a well
     drained one.
 (v) Gales of hurricane force are seldom experienced in the United
     Kingdom, but when they occur they can cause tremendous
     damage. During the past 800 years, storms of this intensity
     took place in 1222, 1703 and 1987.

Note. For complete information on climatic factors as affecting
trees, reference should be made to *Pathology of Trees and Shrubs* by
T. R. Peace (1962).

2 *Fire*

(*a*) Although fire can be one of the greatest hazards in forestry
where large areas of trees grow in close contact with each other, it
does not have the same significance in arboriculture.

(*b*) Specimen trees are not normally grown as close to each other as trees are in forestry, and moreover the conditions under which they grow are completely different.

(*c*) Probably the greatest risk of damage by fire to specimen trees is that of inadvertent action by individuals. Examples are the burning of leaves or rubbish too close to a tree, or damage caused by road repairing equipment which is used to melt the surface of tarred roads.

3 *Domestic animals*

(*a*) Farm animals can cause considerable damage to both old and young trees.

(*b*) Horses will gnaw the bark of large trees and they appear to be particularly attracted by elm and beech. Cows will also damage trees which have comparatively soft bark.

(*c*) Cows, sheep and goats will eat the foliage and twigs of trees of all sizes and, in the case of small trees, the results can be fatal. Unringed pigs will also gnaw the bark of trees and will root up any small ones, if they feel so inclined.

(*d*) The only effective method of protecting trees from farm animals is the provision of fencing or tree guards, and examples are shown in Figure 4.2.

4 *Wild animals*

(*a*) *Rabbits*
  (i) Although the rabbit population in this country was largely decimated by *myxomatosis*, which first occurred in 1953, rabbits have now re-established themselves and are common in many districts.
  (ii) Rabbits will eat the leading shoots and side branches of young trees, and will also gnaw the bark of large trees in hard weather. Ash and beech are particularly susceptible to attack.
  (iii) Control includes shooting, trapping and gassing while protection can be provided by rabbit fencing, tree shelters, spiral plastic tree guards and sleeves of wire or plastic netting.

(*b*) *Hares*
  (i) Hares can cause somewhat similar damage to rabbits. The control of hares is, however, more difficult on account of their

more nomadic habits and their ability to jump a fence which would be a deterrent to rabbits.

(ii) Control may be by shooting, hunting or coursing, the protective measures being similar to those taken against rabbits.

(c) *Squirrels*

(i) The grey squirrel (*Sciurus carolinensis*) is to be found in many parts of this country and has readily adapted itself to life in towns. The red squirrel (*S. vulgaris*) has unfortunately almost disappeared and is now seldom seen.

(ii) Grey squirrels are very destructive and are responsible for killing young birds, destroying eggs, damaging fruit and bulbs and stripping bark from trees. In this last activity they appear to prefer sycamore, beech, ash, and hornbeam, although other species may be damaged. For the most part they direct their activities against trees during the late spring.

(iii) Control is by shooting and trapping, although both these methods are difficult in urban areas.

(iv) Full details regarding grey squirrels are given in Forestry Commission Leaflet No. 56 – *Grey Squirrel Control*, 1980.

(d) *Deer*

(i) Except in parks where herds are artificially maintained, deer are seldom found in urban areas.

(ii) However, in woodland areas on the outskirts of towns, deer are not uncommon, although seldom seen during the hours of daylight. In such surroundings the species concerned are usually Roe, Fallow and occasionally Muntjac.

(iii) Damage consists mainly of fraying or barking the stems of small trees, and browsing the young shoots and leaves of trees and shrubs.

(iv) Control is by shooting, which is covered by The Deer Act 1963 as amended by The Deer Act 1980. A limited amount of protection is provided by tree shelters.

(e) *Mice and voles*

(i) Damage is chiefly confined to gnawing the bark of young trees and eating seeds and buds.

(ii) Although damage can be considerable in forestry where small trees are planted as a matter of course, the larger-sized trees used in urban plantings are usually disregarded.

(iii) Control is by trapping or poisoning, and protection can be provided by spiral plastic tree guards and, to a lesser extent, by tree shelters.

(*f*) *Moles*
  (i) Moles can cause damage to young trees by 'lifting' them during the first year or two of their lives.
  (ii) They can also become a source of trouble in nurseries.
  (iii) Control is by trapping or the use of repellent material in their runs.

5 *Birds*

(*a*) Although in forestry attempts have been made to classify birds under the headings of definitely useful, definitely harmful, and partly useful and partly harmful, the position in the case or arboriculture is very different.

(*b*) In urban areas there are probably only two birds which can be considered as likely to cause damage to trees, and these are pigeons and starlings. Even if a case can be made against other birds, it is very unlikely that action would be taken against them.

(*c*) Starlings can cause considerable damage to plantations and small groups of trees, not only on account of their droppings but also the weight of birds perching on branches.

(*d*) Protection takes the form of fireworks, sulphur smoke and shooting.

6 *Human beings*

(*a*) Human beings are capable of causing more damage to trees than any other agent.

(*b*) It is not necessary to dwell on this unfortunate state of affairs, except to observe that it is unlikely that vandalism will end without increased education, in its widest sense.

7 *Vehicles and traffic*

(*a*) With the increasing volume and speed of traffic, injuries to street and roadside trees are inevitable.

(*b*) In the case of large trees, these injuries usually take the form of damage to the bole with loss of bark and frequently the removal of pieces of the stem. Small trees may be broken off or uprooted.

(*c*) Damage can also be caused to trees in car parks and by the passing of lorries with high loads which damage overhanging branches.

(*d*) In certain cases some protection can be afforded by tree guards, but this is not always practical. Ultimately the best protection is the planting of trees in more suitable positions, whereby the risk of damage is reduced.

8 *Road excavation*

(*a*) Road excavations are carried out for a number of purposes, including the laying and repair of sewers, gas and water mains, electricity mains and telephone cables.

(*b*) Work of this kind invariably consists of digging trenches at varying depths and often in close proximity to roadside trees. As a result two forms of injury may occur:
  (i) Damage to the roots which may affect the general health of the tree, weaken its stability or permit the entry of fungi.
 (ii) Damage to the bole or branches caused by the machinery and equipment used either in excavating or in subsequent work.

(*c*) It is almost impossible to provide any form of protection in these cases.

(*d*) Further information can be obtained from Forestry Commission Arboriculture Research Note 36/85/TRL – *Tree Roots and Underground Pipes*.

9 *Poisoning*

Trees may be poisoned in various ways, and some of these are considered in this section.

(*a*) *Domestic gas*
The leakage of gas from underground pipes can prove fatal to any trees whose roots are in close proximity.

(*b*) *Sewage*
  (i) Under certain conditions trees may suffer from sewage poisoning. This can occur when trees are growing near a cesspool and their roots penetrate into it. Sewage poisoning can also occur where the roots come into contact with the discharge of untreated effluent.
 (ii) A somewhat similar form of poisoning can be caused when farmyard manure is stacked under or in close proximity to trees. In such cases death inevitably occurs.
(iii) In such cases the only form of protection is prevention.

(c) *Toxic substances*
 (i) Trees can also be seriously affected or killed by certain poisonous materials which may be allowed to seep into the surrounding soil or which are inadvertently emptied on to the ground.
 (ii) Examples of these are creosote, oil, petrol, road salt, herbicides, sheep dip liquids and so on.
 (iii) As in other cases, prevention is the best form of protection.

(d) *Poisonous fumes*
Trees can be seriously affected by the discharge of poisonous fumes and dust into the atmosphere. This may occur in and near industrial areas.

Note. Full details of damage caused to trees by poisonous substances will be found in *Pathology of Trees and Shrubs* by T. R. Peace (1962).

B *Protection*

1 *Indirect*

(a) *Inspections and maintenance*
 (i) The regular inspection of trees is an essential step towards their protection. Some forms of injury can be arrested in the early stages provided that it is found in good time.
 (ii) Closely connected with regular inspections is regular maintenance. The adjustment of a tree tie, the timely repair of a tree guard or the removal of a damaged branch will often prevent more serious damage occurring.

(b) *Supervision and management*
Efficient inspection and maintenance will depend to a large extent on proper supervision and management.

(c) *Education*
 (i) Ultimately, the best form of protection against vandalism is the education of the public so that they appreciate the true worth of trees and are encouraged to take a real interest in them.
 (ii) This is undoubtedly a long-term undertaking, but as schools take an increasing interest in nature and wild life, so the prospects improve.

2 Direct

(a) General

The protection of trees against their enemies and misfortunes is never an easy task. In some cases protective action has already been suggested under the sections which describe the damage. Under this heading some of the more general forms of protection are considered.

(b) Tree guards

The following points should be taken into account when deciding which type of guard should be used:

(i) The cause of the damage: this may be due to rabbits, domestic animals, etc.
(ii) The size of the trees to be protected.
(iii) The site on which the trees are growing: this may be a roadside, in a public park, on agricultural land or elsewhere.

(i) *Sleeves*
  (1) These can be made from galvanized wire or heavy-duty plastic netting.
  (2) For small single trees, they should not be less than 60 cm (2 ft) high, supported by two light stakes, as in Figure 4.1(a).
  (3) More permanent protection can be provided by using four corner stakes, each about 90 cm (3 ft) × 2.5 cm (1 in.) × 2.5 cm, as in Figure 4.1(b).

(ii) *Plastic guards*
  (1) These are available either as spiral guards or as split tube guards.
  (2) Spiral guards are obtainable in lengths (when placed in position) of 46 cm (1 ft 6 in.), 60 cm (2 ft) and 75 cm (2 ft 6 in.) But are suitable only for trees with a stem not exceeding 4 cm ($1\frac{1}{2}$ in.) in diameter. The guards are wound around the tree and are supplied in several colours (see Figure 4.1(c)).
  (3) Split tube guards can be purchased in lengths of from 60 cm (2 ft) to 70 cm (2 ft 3 in.) and from 5 cm (2 in.) to 7 cm ($2\frac{3}{4}$ in.) in diameter. Several colours are available. To attach them to the tree, they are opened along the vertical line and then allowed to encircle the stem, as in Figure 4.1(d).

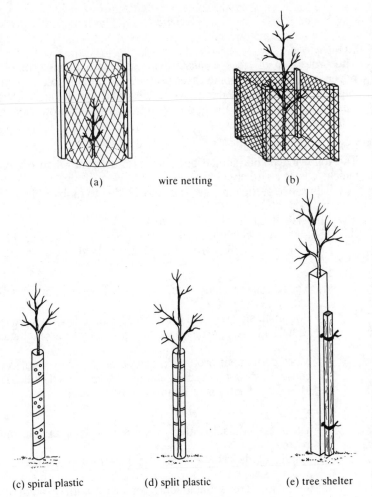

(a)   wire netting   (b)

(c) spiral plastic   (d) split plastic   (e) tree shelter

Figure 4.1 Protection of young trees

(iii) *Tree shelters*
    (1) Tree shelters are tubes of strong plastic which are designed for the protection of small trees at the time of planting.
    (2) They are available with square, triangular, round or hexagonal cross sections. The square pattern measures about 8 cm ($3\frac{1}{4}$ in.) × 8 cm.

(3) Shelters are obtainable in heights of 60 cm (2 ft), 150 cm (4 ft 11 in.) and 180 cm (5 ft 11 in.) with a colour range of brown, green, white or translucent.

(4) Each shelter must be attached to a stake so as to hold it in position, a plastic clip or tying wire usually being included with the shelter, as in Figure 4.1(e).

(5) Wider shelters than those referred to above, are available under the proprietary name of 'Gro-Cones', which are wider at the top than at the bottom. The maximum top diameter that is available is 20 cm (8 in.).

(iv) *Welded wire mesh guards*

(1) This type of guard gives protection from rabbits, hares, deer and farm animals, while the addition of small mesh netting around the lower part of the guard will substantially reduce damage by mice and voles.

(2) They are available in stock sizes of 150 cm (4 ft 11 in.) and 180 cm (5 ft 11 in.) in height and 22.5 cm (9 in.) in diameter, but other sizes can be made to order. The rectangular mesh measures 7.5 cm × 2.5 cm (3 in. × 1 in.) as in Figure 4.2(a).

(v) *Iron guards*

(1) These are constructed of flat iron strips or iron rods, held together by three or four encircling bands. The tops of the uprights are turned outwards and may be pointed.

(2) Guards of this kind can generally be made by a blacksmith or agricultural engineer (see Figure 4.2(b)).

(vi) *Timber guards*

(1) There are several types of these which can be divided into three basic patterns:
Those with vertical protecting framework
Those with horizontal framework
Those modelled on a post-and-rail fence.

(2) The vertical and horizontal patterns rely on height and close spacing of the framework to give protection from deer and farm animals as in Figure 4.2(c).

(3) The fence pattern protects the tree by preventing animals from getting close enough to damage it. Additional protection can be given by the addition of barbed wire or netting, if necessary.

(c) *Tree stakes*

Tree stakes provide protection against wind and to some extent against damage by the public. Staking is dealt with in Chapter II.

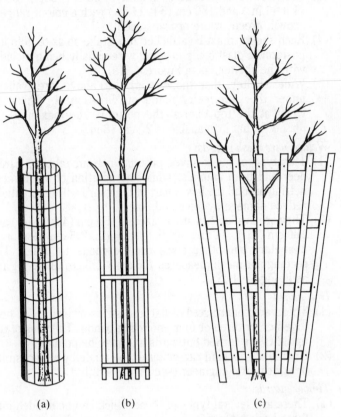

Figure 4.2 Tree guards (a) wire mesh; (b) iron; (c) timber

(d) Screens

Small tender trees and shrubs can be protected from frost and cold drying winds by means of screens of hessian, canvas or sacking.

Further information can be obtained from Forestry Commission Arboricultural Leaflet No. 10 – *Individual Tree Protection*, and from *Woodlands* (pages 95–8), published by the British Trust for Conservation Volunteers.

CHAPTER V

Diseases and Pests

(*a*) This chapter is divided into three sections, the first dealing with fungi and fungal diseases, the second with bacteria and viruses and the third with insects.

(*b*) In 1977, the European Economic Community (EEC) drew up a Plant Health Directive which, amongst other matters, deals with the following points:

(i) The provision of plant certificates for imported plants.

(ii) The action which should be taken to prevent the introduction of between 30 and 40 pests and diseases, with certain special provisions as to the United Kingdom.

(iii) Special emergency measures which should be taken to deal with unforeseen circumstances.

(*c*) The most important tree species to which the Directive applies are, as regards the United Kingdom: *Abies* (silver fir), *Larix* (larch), *Picea* (spruce), *Pinus* (pine), *Pseudotsuga* (Douglas fir), *Tsuga* (hemlock), *Castanea* (sweet chestnut), *Populus* (poplar), Quercus (oak) and *Ulmus*(elm).

(*d*) Further information on these matters can be found in Forestry Commission Forest Record No. 116, *The EEC Plant Health Directive and British Forestry*.

1 *Fungi*

(*a*) In this section, fungi are considered under two main headings:

(i) Those which primarily attack timber, bark or roots.

(ii) Those which affect foliage or cause damage in forest nurseries. These are considered under one heading as there is not always a clear dividing line between them.

In both cases fungi are listed alphabetically under their usual English names.

(*b*) A full account of the diseases of trees which are caused by fungi, bacteria and viruses will be found in *Diseases of Forest and Ornamental Trees* by D. H. Phillips and D. A. Burdekin (The Macmillan Press Ltd, 1982).

A. FUNGI PRIMARILY ATTACKING TIMBER, BARK OR ROOTS

Fungus	Description	Species of tree attacked	Damage	Remarks
1. Anthracnose disease *Gnomonia platani*	Microscopic spores are produced which are spread by wind and rain.	London plane (*Platanus acerifolia*).	Buds fail to open; twigs are girdled and turn orange-brown; shoots wilt and die in spring; leaves turn brown and black.	Healthy trees usually recover. No means of control. See F.C. Arboriculture Research Note 46/83/PATH.
2. Ash canker *Nectria galligena*	Eruptions on the bark, roughly circular in shape up to about 7.5 cm (3 in.) diameter.	Ash.	Serious damage to trunk of poles, which reduces their value to that of firewood.	Ash canker can also be caused by bacterium *Pseudomonas savastanoi*.
3. Beech bark disease *Cryptococcus fagisuga* (syn. *Cryptococcus fagi*, *Nectria coccinea*)	This disease is caused by the Felted Beech Scale (see Section 3D.23 of this chapter) and the fungus *Nectria coccinea*. After infection by the Beech Scale, the fungi enter mainly through the resulting wounds, and patches of bark die. Subsequently, sap oozes like black slime from these areas.	Beech.	After a few years infected trees die and the main stem may break off at a height of 3–5 m (10–15 ft). This is known as 'Beech snap' and is due to other decay organisms and wood-boring beetles.	There is some evidence to suggest that outbreaks of the disease are more likely on downland sites in southern England, on slopes, in pure beech stands and after a heavy thinning. See F.C. Bulletin No. 69, *Beech Bark Disease*.
4. Beef steak fungus *Fistulina hepatica*	Fructification appears in early autumn; at first soft creamy juicy lumps, but soon becomes tongue-shaped. When mature, is deep purplish-red or brown on upper surface and creamy white below. Flesh reddish and marbled.	Chiefly oak and Spanish chestnut, but can attack some other broadleaved trees. Seldom fructifies on young trees, but is found typically on old park oaks, especially if pollarded. Does not attack conifers.	In later stages of attack produces heart rot, but little damage is done to vigorously growing trees.	Spores enter through wounds, and spread to heartwood. Discoloration produces 'brown oak', which may enhance value of tree. Fructification edible. See F.C. Arboricultural Leaflet No. 5.
5. Birch polyphore *Piptoporus betulinus* (syn. *Polyporus betulinus*)	A hoof-shaped fructification, the upper surface being a rather shiny greyish-white, the flesh being thick and white. It is usually found growing on birch trees several feet above ground level.	Birch.	Entering by wounds, the fungus turns the wood brown. In due course the wood is reduced to state of complete rottenness.	Very common on birch throughout the country. See F.C. Arboricultural Leaflet No. 5.
6. Bleeding canker *Phytophora cactorum*	Invisible to the naked eye.	Horse chestnut, lime and apple.	Red, black or light brown sticky liquid seeps out of areas of dying bark. In time tree is girdled and dies.	This fungus can also cause Root Rot as described under No. 20. See F.C. Arboricultural Leaflet No. 8.

	ash, elder, beech, cherry, hazel, hornbeam, lime, sycamore, and fruit trees generally. Probably commonest on ash. Does not attack conifers.	followed by white pustules in late summer and autumn. As attack develops a sunken black wound or canker occurs, which is surrounded by irregularly broken or projecting bark.	only for firewood. Even one large canker will greatly depreciate the value of a tree.	avoid damaging the bark of those trees which are to be left.
8. Brown cubical butt-rot or dry crumbly rot *Phaeolus schweinitzii* (syn. *Polyporus schweinitzii*)	Most coniferous trees but especially Douglas fir, Sitka spruce and larch. Has also been found on oak and cherry.	Fructification bracket-shaped, dark rusty brown in colour, except for growing margin, which is tawny or yellow. Upper surface hairy or velvety. Flesh soft and spongy, becoming fragile on drying. Fructifications invariably occur near ground level, often on roots, and half hidden by soil or leaves.	Infection occurs through roots. Fungus spreads to heartwood, producing a reddish-brown rot, which tends to divide the timber into small cubes. Infected timber produces a sweet/sour resinous smell.	Fungus may grow on a tree for many years before fructification produced. May cause considerable damage. Conifers planted on old hardwood sites are especially subject to attack. See F.C. Leaflet No. 79.
9. Chestnut blight *Endothia parasitica*	Sweet or Spanish chestnut (*Castanea sativa*).	Small yellow, orange, or reddish-brown pustules, about the size of a pin head, appear in very large numbers on the bark.	The fungus kills an area of bark on the main stem of a tree or coppice pole, which very soon girdles the tree and kills that portion above the point of attack.	This fungus has caused tremendous damage to chestnut in the United States of America and Italy. If it became established in this country it is likely that the growing of chestnut would no longer be an economic proposition.
10. Conifer heart-rot or white pocket rot *Heterobasidion annosum* (syn. *Fomes annosus*)	Most conifers, but more seriously Douglas fir, larch, Norway spruce, Western red cedar (*T. plicata*), Western hemlock (*Tsuga heterophylla*). Two of the more resistant conifers are Scots and Corsican pine. Has been known to attack broadleaved trees, but this is uncommon. Stumps of freshly felled trees should be treated with 2% solution of urea.	Fructification can be flat and stalkless or bracket-shaped. The former occur on roots where they are exposed to dim light, e.g. in rabbit holes, or on upper side of large roots which are just below ground level. Brackets occur at base of trunk. Upper surface of fructification is reddish-brown with white margin. Underside white or yellow, perforated by small brown marginal holes.	Infection through roots, rot spreading upwards through heartwood. Timber turns pink and rotten, and turns to tinder if dry or slime if wet. In advanced stages, trees become 'foxed' or 'pumped'. Small trees under 10 years old may be attacked and killed. Scots and Corsican pine are most susceptible to damage at this age.	Can cause serious damage. Appears more prevalent on sites which are too dry or too wet for species concerned, and in a first crop planted on arable land. May be detected by fructification, swelling at base of tree, tapping trunk with stick, use of Pressler borer, or examination of felled or windblown trees. See F.C. Leaflets Nos. 5 and 79.
11. Conifer red-rot *Trametes pini*	Conifers.	Varies in appearance from a thin shell-like bracket to a thick hoof-shaped form. At first dark rusty brown in colour, becoming almost black. Young fructifications have golden yellow margin.	Causes extensive rotting of the heartwood. Infection through wounds.	Although frequently found on the Continent, is uncommon in this country.

Fungus	Description	Species of tree attacked	Damage	Remarks
12. Coral spot *Nectria cinnabarina*	Fructification appears as bright coral red spots. Being both parasitic and saprophytic, it is found on both dead and living branches or stems.	*Parasitically* Apple, horse chestnut, lime, sycamore. *Saprophytically* Elm, hazel, lime, poplar, sycamore.	Causes death of individual branches.	Although very common, is not serious.
13. Coryneum canker *Seridium cardinale* (syn. *Coryneum cardinale*)	Very small dark blisters the size of a pin head.	Monterey cypress.	The fungus enters cracks in the bark and slowly kills the cambium. Foliage turns yellow and brown and falls off. In time, the whole tree may be killed.	First reported in 1969 and now the commonest cause of death in Monterey cypress. Control is by cutting out the infected parts or felling if seriously damaged. See F.C. Arboriculture Research Note 39/84/PATH.
14. Elm disease *Ceratocystis ulmi*	Fructifications minute black bodies, about 1 mm in length. Only visible to naked eye when seen in a mass.	All species of elm grown in this country; also *Zelkova*, an uncommon ornamental tree closely related to the elm.	The disease is mainly spread by the Elm Bark Beetles (*Scolytus spp.*). After the spores have been introduced by feeding beetles into the sap stream, blockages occur, and branch affected dies. Ultimately death of whole tree occurs.	Indications that a tree is affected includes yellowing of the foliage on individual branches in midsummer. Occasional presence of brown spots if twigs cut across; the turning-down of the end of young twigs to form 'shepherds' crooks'. See F.C. Bulletin No. 60, *Research on Dutch Elm Disease in Europe*. *Note*: This fungus could also be included in the second category of fungi, i.e. those attacking foliage.
15. Elm heart-rot *Rigidoporus ulmarius* (syn. *Fomes ulmarius*)	Fructification thick and bracket-shaped. Greyish-green above, and pink below, the underside later turning reddish-brown. May appear at base of tree, or inside hollow elms.	Elms.	The fungus, which enters through wounds, attacks the heartwood, and may lead to complete destruction of the centre of the tree, leaving a thin shell.	The principal cause of heart-rot in elm. Noticeably common in parkland trees which have been damaged by deer. Infected trees should be felled. See F.C. Arboricultural Leaflet No. 5.
16. Group dying of conifers *Rhizina undulata*	Round flattened chestnut-coloured fructifications about two inches in diameter. Produced on soil surface of fire sites.	Chiefly Sitka spruce but also Norway spruce, Corsican and lodgepole pines, European and Japanese larch.	Death of groups of conifers.	Occurs on the sites of fires which have been lit in plantations. Prohibition of such fires is essential. See F.C. Bulletin No. 14.

17. Honey fungus *Armillaria mellea* (syn. *Armillariella mellea*)	Fructification is a brown or honey-coloured toadstool, which often appears in clumps. Upper surface flecked with darker brown scales, under surface made up of 'gills' radiating from the centre to the outer circumference. About 5 mm (¼ in.) from top of the stalk is a ragged ring or annulus.	All conifers and most broadleaved trees. Sitka spruce especially vulnerable on old hardwood sites. Most susceptible species Corsican and Scots pine. Sitka spruce, and Western hemlock. Larch is resistant, and Douglas fir most resistant. Norway spruce is resistant when young, but less resistant when old.	Fungus spreads either by spores or by black underground growths similar to boot laces in appearance, and known as 'rhizomorphs'. These enter through roots, and form mass of felted white mycelium, which frequently kills the tree.	When deaths have occurred through this fungus, and beating up is needed, use a more resistant species. See F.C. Arboricultural Leaflet No. 2.
18. Larch canker *Lachnellula willkommii* (syn. *Trichoscyphella willkommii*)	Fructifications are very small, somewhat similar to a champagne glass in shape, the top being about 5 mm (¼ in) in diameter. Top concave surface yellow or orange, rim and under-surface white. In due course a canker is usually formed, the centre being depressed and resinous, and the outer edges black and broken.	European larch and Western American larch. Only in extremely rare cases have Japanese and hybrid larch been attacked.	The chief damage is caused by the presence of the cankers. In its endeavours to overcome the attack, the tree increases its growth around the canker, so that contortion of the stem occurs, rendering it unfit for conversion into timber.	Considerable doubt seems to exist as to the exact cause of this disease. It appears to be closely associated with frost. There is also some evidence to show that it may enter through dead branches. See F.C. Bulletin No. 14.
19. Phomopsis disease of conifers *Phomopsis pseudotsugae*	Fructifications are small and black, and are found singly or in small groups just beneath the bark, or projecting through cracks in the bark. They are very difficult to see with the naked eye. In damp weather in winter and spring large quantities of white spores are produced.	Principally Douglas fir, but occasionally Japanese larch and other conifers.	Three kinds of damage occur: (a) Die-back of leading or lateral shoots. (b) Girdling of the stem (this spreads from the die-back of a lateral shoot). (c) Canker of the stem.	Die-back of leading shoots may result in serious damage in nurseries and young plantations. Girdling may kill young trees 6 to 8 years old, while canker occurs particularly on trees 15 to 20 years old.
20. Phytophthora Root Rot *Phytophthora* spp.	Invisible to the naked eye.	Ash, beech, Horse and Sweet chestnut, Lawson cypress, lime *Nothofagus*, yew and others.	Attacks root system which affects foliage. Leaves become small and scattered, and foliage turns yellow. Bark may die at and above ground level.	See F.C. Arboricultural Leaflet No. 8.
21. Pine blister or Resin-top disease *Peridermium pini*	Fructification appears as orange-yellow tufts which break through the bark of trees in midsummer. Spores are then discharged and the fructification turns white.	Scots pine is the usual subject of attack, but any other species of pine may be affected.	The fungus develops underneath the bark, attacking the bast and cambium. If it girdles the stem, the portion of the tree above the infection will die.	Trees from 15 to 20 years seem to suffer most. Correct thinning will do much to prevent the disease. Any infected trees should be cut out and removed. See F.C. Bulletin No. 14.

Fungus	Description	Species of tree attacked	Damage	Remarks
22. Poplar canker *Dothichiza populea*	Dead patches occur on stem often at base of twigs or around wounds. Attacked area often 'sunk' below level of stem.	Poplars but *P. serotina var. erecta* and *P. gelrica* seldom attacked.	Die-back occurs particularly in poplars which have been overcrowded in the nursery or badly planted.	Avoid overcrowding in nursery and improper planting. Prune nursery stock only in summer.
23. Saddleback fungus, or Dryad's saddle *Polyporus squamosus*	Fructification a fan-shaped bracket up to 60 cm (24 in.) across. Upper surface pale fawn in colour, with dark brown scales which appear as flecks. Under-surface creamy white. Fructifications decay very quickly.	Probably it is most commonly found on elm, but most broadleaved trees are attacked. Conifers are not affected.	Infection occurs through a branch wound; in the case of elms, where a limb has fallen. The fungus penetrates to the centre of the tree, and heart-rot ensues.	Infected trees should be felled and removed. See F.C. Arboricultural Leaflet No. 5.
24. Sooty bark disease of sycamore *Cryptostroma corticale*	Fructifications appear to be minute black columns about 1 mm long, which in due course produce a large mass of blackish-brown spores having the appearance of soot—hence the common name.	Sycamore.	The fungus works beneath the bark, which ultimately causes the death of the infected branch or stem.	This disease was first noticed in Wanstead Park, London, in 1945. After an interval, an exceptionally severe outbreak occurred in 1976 on sites up to 160 miles from London. See F.C. Arboricultural Leaflet No. 3.
25. Sulphur polyphore *Laetiporus sulphureus* (syn. *Polyporus sulphureus*)	Fructifications are orange above and pale sulphur yellow below. Bracket-shaped with a wavy margin, the flesh is soft and soon decomposes.	Broadleaved and coniferous species are attacked. Oak and cherry are two of the commonest, but many other trees may be affected.	The fungus enters through wounds, and spreads to the heart-wood, causing butt-rot.	A very common cause of heart-rot in old oaks. See F.C. Arboricultural Leaflet No. 5.
26. White pine blister *Cronartium ribicola*	Swollen canker-like growth on branches and main stem, on which white blisters containing orange spores are produced in spring and early summer.	All five-needle pines.	Small shoots are completely girdled and cankers formed on the main stems. In many cases the tree is killed.	Alternate host on currants (*Ribes spp.*). Effect of this fungus has eliminated five-needled pines, especially Weymouth pine (*P. strobus*) in British forestry.
27. White rot *Ganoderma applanatum*	Bracket fructification, grey-brown above with whitish under-surface.	Several broadleaved species, but very common on beech.	Infection generally enters through wounds and broken branches. Extensive heart rot occurs, which ultimately renders the tree valueless for timber.	If the whitish under-surface is marked with a knife, a red line will appear. See F.C. Arboricultural Leaflet No. 5. which also covers other species of *Ganoderma*.

B. FUNGI ATTACKING FOLIAGE AND NURSERY STOCK

28. Beech seedling blight *Phytophthora cactorum*	Invisible to the naked eye.	Beech seedlings.	Attacks the cotyledon and first true leaves, which turn brown. Seedling frequently dies.	The spread of the disease can be controlled by spraying with a fungicide as used on Potato blight. See F.C. Arboricultural Leaflet No. 8.
29. Corsican pine die-back *Gremmeniella abietina* (syn. *Brunchorstia destruens*)	Black fructifications (*pycnidia*) up to 3 mm (⅛ in.) diameter on dead needles, buds and shoots, especially from November to March.	Corsican pine; occasionally Austrian, Scots and Maritime pines.	Die-back of shoots and loss of needles, thus affecting growth and increment. Symptoms may be confused with exposure.	Avoid planting Corsican pine on sites which sun does not reach, e.g. northern slopes and deep valleys or where humidity is high or shade excessive.
30. Douglas fir leaf cast *Rhabdocline pseudotsugae*	Fructifications appear as bright brown, slightly protruding patches on the underside of the needles.	Douglas fir. The Oregon or Green variety is the most resistant; the Colorado or Blue Douglas the most susceptible.	Defoliation, since the infected leaves fall at all times of year, with consequent loss of growth.	Infected nursery stock should be destroyed. Where attacks have occurred previously, spraying with soap and Bordeaux mixture from mid-April to mid-June may be successful.
31. Douglas fir leaf cast *Phaeocryptopus gaumanii*	Minute black patches appear in irregular lines on the needles.	The coastal type of Douglas is least affected.	Defoliation and loss of growth.	Little can be done other than destruction of infected trees.
32. Grey mould *Botrytis cinerea*	Grey brown 'misty' mould on needles and leaves.	All species, but conifers worse than broadleaved, especially Sitka spruce, Douglas fir, Western hemlock, *Cupressus* and *Sequoia*.	This fungus can cause considerable damage in forest nurseries by killing young seedlings and weakening others through loss of needles.	Control by spraying with Bordeaux mixture or more modern counterparts.
33. Larch leaf cast *Meria laricis*	The fungus appears as clusters of minute white spots on the lower, but sometimes upper, surface of the needles. A more certain diagnosis can be made by staining the needle with aniline blue, which causes the spots to turn black.	European and American larch. There is no record of Japanese larch having been attacked.	The needles turn brown in early May, resembling frost damage, but the following differences should be noted: (a) Frost kills whole of needle at once. (b) *Meria* browns middle or end. (c) Needles killed by frost tend to remain on twig throughout the season. (d) Needles killed by *Meria* fall as soon as totally brown. (e) Needles killed by *Meria* are less shrivelled than those killed by frost.	Damp seasons favour the disease. 2-year seedlings tend to suffer worst. *Meria* plus frost may cause heavy losses. Spraying with sulphur spray consisting of amberene, suisol, and liver of sulphur is effective. See F.C. Bulletin No. 43.

Fungus	Description	Species of tree attacked	Damage	Remarks
34. Oak mildew *Microsphaera alphitoides*	The fungus, consisting of minute white fructifications, appears as typical grey 'mildew'.	Pedunculate and sessile oaks, the former suffering most. Also recorded on beech and sweet chestnut.	Interference with the natural functions of the leaf, and often defoliation. Particularly bad in conjunction with attacks by oak leaf roller moth (*q.v.*). Successive attacks may prove fatal. Most serious in nurseries.	Control by spraying with sulphur spray (colloidal sulphur wash) or dry dusting with sulphur. Its natural enemies include a parasitic fungus (*Cicinnobolus spp.*) and larvae of cecidomyid fly. See F.C. Bulletin No. 43.
35. Oak seedling disease *Rosellinia quercina*	No fructification is apparent on a growing seedling. The leaves at first become pale and then wither; the upper leaves wither first, and then the lower ones. This is particularly noticeable in a wet season. On lifting an infected seedling, some small black bodies about the size of a pin head will be found on the tap-root.	Oak seedlings 1–3 years old.	Considerable losses may be caused by this fungus, which proves fatal to young seedlings. The fungus attacks through the roots.	All infected plants should be removed and burnt, and seedbeds on which oak have been grown should be used for other species.
36. Pine needle cast *Lophodermium seditiosum* (syn. *Lophodermium pinastri*)	Small black spots on needles which later turn brown and fall immaturely.	Scots pine and Corsican pine.	Defoliation through early leaf fall. Most serious amongst nursery stock.	Not a serious disease in plantations but can cause losses in nurseries. Spray with Bordeaux mixture or other suitable fungicides. See F.C. Bulletin No. 43.
37. Pine twisting rust *Melampsora pinitorqua*	Bright yellow areas of fruiting bodies like wet sawdust, produced in June on young shoots.	Scots pine with aspen poplar as alternate host. Other pines rarely attacked.	Distortion and subsequent death of current year's shoots and foliage.	Cut out all aspen near Scots pine plantations, or plant Corsican pine in aspen areas.
38. Keithia disease *Didymascella thujina* (syn. *Keithia thujina*)	Fructifications appear as minute black spots on the underside of the leaves. The fungus enters through the leaves.	All varieties of *Thuya* are liable to attack, but from a forester's point of view the most important is the Western red cedar (*Thuya plicata*).	Although the most damage is caused to seedlings, young transplants can also be attacked. Results are often fatal in the case of seedlings.	Spraying with sulphur wash, or cycloheximide potassium permanganate is recommended. Badly infected stock may have to be burnt. In some cases the cutting back of the foliage, followed by a dressing of farmyard manure, will produce fresh healthy leaves. See F.C. Bulletin No. 14.
39. Verticillium wilt (*Verticillium dahliae*)	A micro-fungus which cannot be easily seen or identified.	Maples, limes and some other ornamental trees including the sumach and Judas tree.	In its worst form, foliage wilts, turns brown, and death follows.	A serious disease of young trees which generally occurs in the nursery. See F.C. Arboricultural Leaflet No. 9.

2 *Bacteria and viruses*

(*a*) *Bacteria*

Three diseases caused by bacteria – Watermark Disease of cricket bat willows, poplar canker and Fire Blight are described in the table on page 60.

(*b*) *Viruses*

(i) Although a considerable amount of research has been carried out on the virus infection of farm crops and fruit trees, work on trees and shrubs has started only comparatively recently.

(ii) The viruses which affect plants consist of minute particles which can spread into all the tissues of the plant concerned. The symptoms of virus infection can occur in the foliage, often causing discoloration, or in the stems or branches, producing cankers, patches of dead bark or grooves.

(iii) There are in addition 'virus-like diseases', one group of which comprises into two main types:
Rickettsia-like organisms (R.L.O.s)
Mycoplasma-like organisms (M.L.O.s)

(iv) Detailed information will be found in Arboricultural Leaflet No. 4, *Virus and Virus-like Diseases of Trees*, issued by the Forestry Commission.

BACTERIAL DISEASES

Disease	Description	Species of tree attacked	Damage	Remarks
1. Watermark disease of cricket-bat willow *Erwinia salicis* (*syn. Bacterium salicis*)	This disease causes die-back of the crown, which is accompanied by a dark staining within the tree. The symptoms are briefly as follows: leaves of infected part whither and turn reddish in April and May; this is accentuated by hot weather occurring in next few weeks. Watery liquid may be discharged from affected shoots or holes made by insects. The timber of infected trees turns dark brown or black; seen in cross-section this may affect only the heartwood, or may appear as concentric circles. Watermark disease is sometimes mistaken for honey fungus (see preceding section of this chapter).	Cricket-bat willow (*Salix alba* var. *coerulea*) White willow (*S. alba*) Crack willow (*S. fragilis*) Sallow (*S. cinerea*)	The timber of the tree is rendered unfit for use. In the case of bat willows this may cause a very serious financial loss to the owner. Incidental damage in the die-back of the crown.	The disease is thought to be spread by small birds, and by insects such as willow sawfly, willow gall midge, and goat moths. The use of infected sets is a more obvious cause. Infected trees should be felled and burnt. Statutory orders have been made for the destruction of infected willows in certain countries. See F.C. Arboricultural Leaflet No. 20.
2. Poplar canker *Pseudomonas syringae* forma *populae*	Small cracks appear in the bark of young twigs. In the spring (especially in April and May) pale brown bacterial slime exudes. The affected twigs may die, but more often a canker develops around the wound. In time the cankers may girdle and kill large branches, and disease becomes noticeable through dead branches, cankers, and discharge of slime.	Many varieties of poplar are affected, and below are given seven of the *resistant* species: *P. alba* *P. berolinensis* *P. canescens* *P. eugenei* *P. gelrica* *P. laevigata* *P. serotina*	Serious reduction in the value of the timber, and in some cases death of the tree.	Cankered trees should be felled and burnt.
3. Fire Blight *Erwinia amylovora*	Bacteria enter through the flowers, which turn black and shrivel. Leaves are then attacked, with cankers forming on twigs and small branches.	Pears, apples and *Crataegus spp.*, especially hawthorn.	Serious die-back and canker that can kill the tree.	Under The Fire Blight Disease Order 1958, trees can be compulsorily felled. See Ministry of Agriculture (ADAS) leaflet 571 (1982).

3 *Insects*

(*a*) The injurious insects which are described in the following table have been divided into five classes, namely:

A. Foliage-destroying insects.
B. Excavating insects other than those which bore into timber.
C. Timber-boring insects.
D. Bark-feeding insects.
E. Root-feeding insects.
F. Gall-making insects.

(*b*) The list of insects given in the table is only a selection of the more important ones, and should not be regarded as exhaustive.

(*c*) Further information on woodland insects will be found in Forestry Commission Handbook No. 1, *Forest Insects*, while drawings of many species are included in R. Neil Chrystal's *Insects of the British Woodlands*, which was first published in 1937.

(*d*) Where measurements of insects are given they refer to a fully-grown specimen.

A. FOLIAGE-DESTROYING INSECTS

Insect	Description	Species of tree attacked	Damage	Remarks
1. Green spruce aphid *Elatobium abietinum* (syn. *Neomyzaphis abietina*)	A minute insect about 1 mm (0.03 in.) long, which exists in both winged and wingless forms. Body green with red eyes.	Various species of spruce, but Sitka spruce is particularly subject to attack.	Insects feed on the needles of the spruce, which ultimately turn brown and die.	The natural enemies of this pest include ladybirds, lace-wing flies, hover flies, tits and tree-creepers.
2. Large larch sawfly *Pristiphora erichsonii* (syn. *Lygaeonematus erichsoni*)	*Larva*: 19 mm (¾ in.) long. Body greyish-green above, with black hairy head. *Adult*: 19 mm (¾ in.). Wing span 25 mm (1 in.). Wings transparent. Head black. Body black with red band around abdomen.	Larch	The caterpillars eat the needles, which may lead to severe defoliation.	This insect caused very serious damage in Cumberland between 1906 and 1912. Since then there have not been any definite outbreaks. Adults appear from late May onwards.
3. Oak leaf roller moth *Tortrix viridana*	*Larva*: 13 mm (½ in.) long. Greyish-green above, with black or brownish-black head. *Adult*: 8 mm (⅓ in.) long. Wing span 19 mm (¾ in.). Fore wings bright green; hind wings grey; all wings have white fringes.	Pedunculate oak (*Q. robur*) and sessile oak (*Q. petraea*), the former being subject to the more serious attacks. Other broadleaved trees adjacent to oaks may also be attacked, particularly the sweet chestnut.	In serious cases, complete defoliation may result, which causes loss of increment and poor seed production. Attacks seldom, if ever, prove fatal, unless there is some additional cause, such as honey fungus (see first section of this chapter).	The natural enemies of this moth include birds such as thrushes, tits, starlings, and finches. Other enemies are ants, earwigs, and ichneumon flies. Adults appear in June and July.
4. Pine beauty moth *Panolis flammea*	*Larva*: 3 mm (⅛ in.) long and grey-green on hatching. Later grey-green with seven stripes lengthways. Finally, 38 mm (1½ in.) long with two orange stripes before pupation. *Adult*: Fore wings yellowish-brown with reddish tinge and cream markings. Hind wings dark grey.	Lodge pale pine.	Heavy defoliation resulting in death of tree	Severe outbreaks in Scotland 1976–77–78. See F.C. Forest Record No. 120.

5. Pine looper moth *Bupalus piniaria*	*Larva:* Green with five white or cream longitudinal stripes. Head green. *Adult:* Wing span 38 mm (1½ in.). Male: dark brown wings with cream patches on wings near body. Antennae are bipectinate ('herringbone'). Female: Orange-brown with darker brown markings. Antennae filiform ('thread-like').	Pines, especially Scots pine and Corsican pine.	Heavy defoliation resulting from the caterpillars feeding on the needles.	Although a serious pest on the Continent, the most serious outbreak in this country occurred at Cannock Chase in 1953. Adults appear in May and June. See F.C. Forest Record, No. 119.
6. Large pine sawfly *Diprion pini*	*Larva:* Up to 25 mm (1 in.) long. Pale green at first, turning brownish-green. 22 legs. Feet black, but turn yellow. *Adult:* Male: 15 mm (⅝ in.) wing span. Body black, but apex reddish. Wings transparent. White spots underside of first segment. Female: 20 mm (⅞ in.) wing span. Body dull yellow: middle of abdomen black.	Pines, especially Austrian, Corsican and Scots pines.	Serious defoliation.	Adult flies normally appear from late April to early June.
7. Spruce sawfly *Gilpinia hercyniae*	*Larva:* Dull green at first, becoming bright green with white stripes. Max length 15–20 mm (⅝–¾ in.)	Spruces, mainly Norway and Sitka owing to their extensive use in forestry.	Serious defoliation causing a general loss of increment.	Of increasing importance on account of the large areas of spruce which have been planted. See F.C. Forest Record No. 117.
8. Vapourer moth *Orgyia antiqua*	*Larva:* 25 mm (1 in.) long. Brown to grey, with distinctive tufts of hair. Head black. *Adult:* Male: Wing span 38 mm (1½ in.). Wings yellowish to brown, hind pair with white crescent-shaped spot. Female. Brown and virtually wingless, grey in colour.	All kinds of trees and shrubs.	Defoliation.	Not normally serious, but sometimes becomes too numerous.

B. EXCAVATING INSECTS OTHER THAN TIMBER BORERS

Insect	Description	Species of tree attacked	Damage	Remarks
9. Ash bud moth *Prays fraxinella* (syn. *Prays curtisella*)	*Larva:* Greenish-grey with black head, 12 mm (½ in.) long. *Adult:* Wing span up to 19 mm (¾ in.). Fore wings whitish-grey with brown patch; hind wings brown.	Ash.	Destruction of the terminal buds, causing, in the case of a leading shoot, forked growth. This can be particularly serious in young plantations.	Any outbreaks in the nursery should be dealt with by cutting out infected shoots. Little can be done in plantations.
10. Douglas fir seed fly *Megastigmus spermotropus*	*Larva:* Small white legless grub. *Adult:* 6 mm (¼ in.) long. Dirty yellow in colour. Female has long sting-like projection for laying eggs. In male this is absent.	Douglas fir.	Grub feeds on seeds of Douglas fir, so that only an empty husk is left.	Control by fumigation of seeds.
11. Larch miner moth *Coleophora laricella*	*Larva:* 6 mm (¼ in.). Reddish-brown with black head. *Adult:* Wing span 10 mm (⅜ in.). Head and wings grey; distinctive marginal fringe to both fore and hind wings, the fringe being deeper on the latter.	Larch.	Damage to needles, caused by caterpillars feeding on them.	The caterpillar builds a case around itself from the remains of damaged needles. Only the head of the caterpillar projects, and when feeding it raises its body and the case in the air.
12. Larch shoot moth *Argyresthia laevigatella* (syn. *Argyresthia atmoriella*)	*Larva:* 6 mm (¼ in.) long. Greenish-yellow body, with black head. *Adult:* 5 mm (⅜ in.) long. Wing span 10 mm (⅜ in.). Fore wings grey and glossy; hind wings darker and not glossy.	European and Japanese larch.	Twigs or shoots of larch are attacked by caterpillars, which burrow into the twig so that it ultimately dies.	Very difficult to control.

13. Common pine shoot beetle *Tomicus piniperda* (syn. *Myelophilus piniperda*)	*Larva*: Small, white, legless, having a wrinkled appearance. *Adult*: 3–5 mm (⅕ in.) long. Head and thorax black, shining but hairy. The wing covers are reddish-brown, with rows of holes running lengthways from front to rear. The second row from the centre outwards does not continue the whole length of the wing, but ends where the cover curves downwards.	Chiefly Scots pine, but other pines, and sometimes larch and spruce, are attacked.	Beetles bore into pith of young shoots, causing shoot to die or be broken off by the wind. Continuous attacks seriously affect increment, crown development is restricted, less seed is produced, and trees may be killed. Beetles also bore under the bark of suppressed or sickly trees, in order to construct breeding galleries.	Control measures include: removing sickly trees by regular thinnings; avoid leaving felled trees in woods; trapping by leaving decoy logs which are later removed and their bark burnt. See F.C. Leaflet No. 3. *Note*: This insect could also be included in Section D – Bark-feeding Insects – but this is only its secondary form of damage.
14. Lesser pine shoot beetle *Tomicus minor* (syn. *Myelophilus minor*)	Similar to the pine shoot beetle (*T. piniperda*), except that the second row of holes in the wing covers continues the whole length of the wing cover. It is very slightly smaller.	Chiefly Scots pine. The beetle is chiefly found in and around Aberdeenshire, but it has also been found in the New Forest.	As for *T. piniperda*.	The breeding galleries of the species are entirely different. The main tunnel of *T. piniperda* runs vertically up the axis of the tree, while *T. minor* runs horizontally across the axis. For further details see F.C. Leaflet No. 3.
15. Pine shoot moth *Rhyacionia buoliana* (syn. *Evetria buoliana*)	*Larva*: Dark brown with black head. *Adult*: Wing span 15–23 mm (⅗–1 in.). Fore wings orange and red with silver lines: hind wings grey. All wings have a fringe of grey hairs.	Pines in the following order of susceptibility: Lodgepole, Scots, Austrian, Corsican.	Destruction of the leading shoot, which is subsequently replaced by a side shoot, resulting in ultimate distortion of the main stem. Such malformation is known as 'posthorn' development.	The species, which were previously included under *Evetria*, have been divided into four genera: *Rhyacionia*, *Blastesthia*, *Petrova* and *Clavigesta*.

C. TIMBER-BORING INSECTS

Insect	Description	Species of tree attacked	Damage	Remarks
16. Goat moth *Cossus cossus*	*Larva:* 75 mm (3 in.) long. Upper surface of body bluish-red; remainder flesh-coloured. Head and first few segments dark brown or black. Larvae produce a strong, unpleasant smell – hence the name *goat* moth. *Adult:* Wing span 90 mm (3½ in.). Reddish-brown head; wings greyish-brown.	Broadleaved species, more particularly oak, ash, willow, and poplar.	The caterpillars bore into the trunk of the tree, causing severe damage to the timber.	Control is very difficult, especially in the case of forest trees.
17. Larch longicorn beetle *Tetropium gabrieli*	*Larva:* 20 mm (¾ in.) long. Brown head and cream body. Three pairs of legs. *Adult:* 20 mm (¾ in.) long. Dark brown or black in colour. Antennae about 90 mm (3½ in.) long.	Larch.	Boring in the sapwood of the tree, which reduces the value of the timber.	Only dying or sickly trees are normally attacked.
18. Large poplar longhorn *Saperda carcharias*	*Larva:* 30 mm (1¼ in.) long. Pale yellow and legless. *Adult:* 30 mm (1¼ in.) long. Yellow to grey in colour, with two bands of lighter yellow across the wing cases. Antennae about 20 mm (¾ in.) long.	Poplar.	Larvae tunnel under the bark, and in addition to damaging the timber may kill the tree by ringing it.	
19. Leopard moth *Zeuzera pyrina*	*Larva:* 50 mm (2 in.) long. Yellowish-white, with dark patches on the head and first and last segments. *Adult:* 75 mm (3 in.) long. Wings white, with black spots.	Broadleaved trees generally.	The timber is damaged by larvae boring, and where small branches are attacked these may subsequently break off at the point of attack.	

D. BARK-FEEDING INSECTS

Insect	Description	Species of tree attacked	Damage	Remarks
20. Ash bark beetle *Leperisinus varius* (syn. *Hylesinus fraxini*)	*Larva:* Small, white, with dark head. *Adult:* 3 mm (⅛ in.) long. Black with grey and black scales.	Ash.	Tunnels immediately beneath the bark, which is not a serious defect from the timber aspect.	
21. Black pine beetle *Hylastes ater*	*Larva:* Very small and white. *Adult:* 5 mm (⅕ in.) long. Black in colour, shining and smooth.	Conifers, especially Scots pine.	The bark and cambium layer of young conifers is attacked at ground level, while the roots are frequently damaged. When young trees are badly injured they turn brown and lose their needles. The beetle breeds in the stumps of roots of dead and dying conifers.	There are six species of hylastes of which the most important are *H. ater* (in southern districts), *H. brunneus* (in northern districts) and *H. cunicularius* (mainly on spruce). For further details see F.C. Leaflet No. 58.
22. Large elm bark beetle *Scolytus scolytus*	*Larva:* 13 mm (½ in.) long, white. *Adult:* 6 mm (¼ in.) long; black shining thorax, with reddish-brown wing cases. *Note:* The adult of the small elm bark beetle (*S. multistriatus*) is similar in size; it is 3 mm (⅛ in.) long.	All species of elm which grow naturally in Great Britain.	This beetle and the small elm bark beetle (*Scolytus multistriatus*) are the chief agents for spreading Elm Disease (see Section 1.A.14 of this chapter). The breeding galleries, which the beetles construct under bark, form ideal locations in which the fructifications of the fungi can develop.	On emergence, beetles which have come into contact with the fungus fly to healthy trees, and feed on young twigs, thus passing on the disease.
23. Felted beech scale *Cryptococcus fagisuga* (syn. *Cryptococcus fagi*)	*Larva:* Minute. Yellow, with three pairs of legs and two antennae. *Adult:* 1 mm (1/25 in.) long. Lemon-coloured. Wingless and legless.	Beech.	The insect feeds on the sap and juices beneath the bark. Badly infected trees are covered with white masses of the insect, giving the bole the appearance of being white-washed.	The precise effect of this insect on the growth of the tree is uncertain. It has been suggested that only sickly trees are affected. See also Section 1.A.3 of this chapter and F.C. Bulletin No. 69.

Insect	Description	Species of tree attacked	Damage	Remarks
24. Large pine weevil *Hylobius abietis*	*Larva*: 15 mm ($\frac{1}{2}$ in.) long. White, with brown head. Body curved and wrinkled, with row of round breathing holes. *Adult*: 15 mm ($\frac{1}{2}$ in.) long, brown to black in colour, the wing cases having an irregular pattern of yellow scales on them. The antennae are at the tip of the snout, and the thickest joint of each leg is notched.	Conifers of any kind, but young broadleaved trees will be attacked if there are no conifers available.	The weevil feeds on the bark of young conifers, and may completely ring a young tree, with fatal results. Less serious attacks will weaken the vigour of young trees.	Favourite breeding places are the stumps of pines, especially Scots and Corsican. Areas which have been burnt are particularly suitable as breeding sites. See F.C. Leaflet No. 58.
25. Banded pine weevil *Pissodes pini* *Pissodes castaneus* (syn. *Pissodes notatus*)	*Larva*: Similar to *Hylobius*, but smaller. *Adult*: 8 mm ($\frac{1}{3}$ in.) long. (*P. castaneus* is rather smaller.) Reddish-brown in colour, with two distinct bands of yellow scales. The antennae are in the middle of the snout, and the largest limbs are *not* notched.	Conifers.	The larvae damage the cambium, and may girdle the stem. Adult weevils attack all parts of a tree.	Young trees up to 10 years old are attacked, and also older trees from 35 to 40 years old.
26. Spruce bark beetle *Ips typographus*	*Larva*: 6 mm ($\frac{1}{4}$ in.) long. Fat, white, and legless, with brown heads. *Adult*: 5 mm ($\frac{1}{5}$ in.) long. Dark brown to black. At the rear the wing cases appar to be notched when seen in profile.	Spruce.	If a heavy infestation occurs, the effect of large numbers of beetles feeding on the bark of a tree may kill it.	The spruce bark beetle is uncommon in this country, but imports of unbarked spruce logs in 1946 are thought to have resulted in the introduction of a small number of beetles. There are several other species of the genus *Ips*.
27. Great spruce bark beetle *Dendroctonus micans*	*Larva*: 6 mm ($\frac{1}{4}$ in.) long. White with brown head. *Adult*: 6–8 mm ($\frac{1}{4}$ in.) long; 3 mm ($\frac{1}{8}$ in.) wide. Brown becoming black when mature. Covered with orange coloured hairs.	Most species of spruce and sometimes Scots pine.	Larvae feed under bark and may eventually girdle and kill the tree. This may take several years.	Signs of attack are: resin running down the trunk and clots of resin and frass at the base. Eggs and larvae are consumed by predator, ***Rhizophagus grandis***. See F.C. Research Information Note No. 128.

E. ROOT-FEEDING INSECTS

Insect	Description	Species of tree attacked	Damage	Remarks
28. Cockchafer *Melolontha melolontha*	*Larva:* 38 mm (1½ in.) long. White and fleshy. Brown head, with powerful jaws. Three pairs of legs. *Adult:* 25 mm (1 in.) long. Head and thorax black; wing covers reddish-brown. A strong flyer.	Young trees of all species.	The larvae gnaw through or girdle the roots of young trees in the nursery. In a dry season this may have fatal results.	There are several species of chafers. Control is difficult. Cultivation in the nursery, hand picking, and encouragement of natural enemies are probably most successful.

F. GALL-MAKING INSECTS

Group	Species	Gall on spruce	Alternative host	Appearance on alternative host	Remarks
29. Spruce-larch	*Adelges viridis*	At base of shoot which it may encircle.	Larch.	White 'wool' on needles, trunk and branches of European larch. Bark exfoliation i.e. flaking off in 'plates', on Japanese and hybrid larch.	See F.C. Handbook No. 2, *Forest Insects*
	Adelges laricis	On, or at end of, shoot.	Larch.	No 'wool'. This adelgid is more harmful to larch than spruce.	
30. Spruce-silver fir	*Adelges nordmannianae* (syn. *A. nusslini*)	Very rare in Britain.	Common silver fir (*Abies alba*). Caucasian fir (*A. nordmanniana*).	'Wool' which occurs at base of needles, on branches and main stem of young trees, is less conspicuous.	See F.C. Handbook No. 2
	Adelges piceae	Not found in Britain.	Most species of silver fir.	Large amount of 'wool' and many small galls are produced on stems and branches.	
31. *Spruce-pine*	*Pineus pini*	Only on Oriental spruce (*Picea orientalis*).	Scots pine and some other pines.	Large quantities of 'wool' are produced.	See F.C. Handbook No. 2.
	Pineus strobi	None.	Weymouth pine (*Pinus strobus*).	'Wool' on bark of branches and main stem.	
32. Spruce-Douglas fir	*Adelges Cooleyi*	Up to 64 mm (2½ in.) long and enclosing the shoot. Very often on Sitka spruce.	Douglas fir, larch, pine or silver fir.	'Wool' on underside of needles, with black specks during the winter.	See F.C. Handbook No. 2

CHAPTER VI

Tree Surgery

During recent years the basic principles of tree surgery have undergone a fundamental change and this is very largely due to the work of Dr A. L. Shigo, K. Vollbrecht and N. Hvass. A full account of the results of their researches will be found in *Tree Biology and Tree Care* by A. L. Shigo, K. Vollbrecht and N. Hvass, published by SITAS, Skovvej 56, 2750 Bellerup, Denmark, and obtainable from Honey Brothers Ltd, New Pond Road, Peasmarsh, Guildford, Surrey GU3 1JR. The following publications by Dr Shigo can be obtained from Shigo and Trees, Associates, 4 Denbow Road, Durham, New Hampshire, USA:
New Tree Biology
Tree Pruning
New Tree Health.
The existence of possible decay and indications of it, are described in two Forestry Commission publications:
Arboriculture Leaflet No. 1 – *External Signs of Decay in Trees*
Leaflet – *The Recognition of Hazardous Trees.*

1 *Pruning*

(a) *The reasons for pruning*
The pruning of trees is carried out for one or more of t. : following reasons:
  (i) To produce a given length of the main stem, or trunk, free from branches.
 (ii) To help the tree to form a satisfactory well-balanced crown.
(iii) To reduce the weight or alter the shape of the crown, if it is too large.
 (iv) To remove diseased, dead or dying branches.
  (v) To reduce leaf surface and in this way decrease the demands made on the root system.
 (vi) To remove damaged branches.

71

(vii) To prevent the branches interfering with overhead cables, buildings, passing traffic, and to take such steps as may be necessary in the interests of safety.

(*b*) *Season for pruning*
  (i) Pruning is usually carried out between October and March when the trees are dormant, the sap is down and when deciduous trees are leafless.
 (ii) However, trees can be pruned at other times of the year, but the following points should be borne in mind:
      (1) The movement of sap becomes very active in the spring and if a tree is cut during this period there will be a considerable discharge of sap. This is known as 'bleeding'.
      (2) Some species are particularly prone to bleeding and these include birch, beech, hornbeam, maple and sycamore.

(*c*) *Pruning young trees*
  (i) In the case of young trees which are intended to grow on into large specimens, probably the most important point is to maintain a straight, unforked main stem and a good leader.
 (ii) Double leaders must be removed as early as possible and if necessary the remaining one must be fastened to a stake to bring it back into the vertical.
(iii) A clean bole is usually desirable and this can be achieved by gradually removing unwanted branches. This should be done before they become too large.
(iv) Any branches which are likely to affect the shape and balance of the tree adversely in the foreseeable future should be removed as soon as it is practical to do so.
 (v) On account of their habit of growth, young conifers usually need to be pruned far less than broadleaved trees. It is, however, most important to deal with double or multi-leadered specimens at an early age.

(*d*) *Pruning of large trees*
  (i) *Removal of large branches*
      (1) Under certain circumstances it may be necessary to remove large branches from a mature or near-mature tree. Such work needs considerable skill, experience and care, and should be carried out only by those who are fully conversant with operations of this kind.
      (2) The branch which is to be removed must be sawn through in such a way that its severance will not cause damage to the tree by tearing the bark. This is known as 'spauling'.

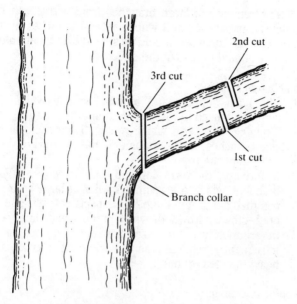

Figure 6.1 Removal of a large branch

(3) To avoid this, the branch should first be undercut, not less
     than 30 cm (12 in.) from the point at which the ultimate
     severance is required. Sawing should continue for about
     half the thickness of the limb or until the saw begins to
     'pinch'. The saw is then removed and a top cut made about
     8 cm (3 in.) in front of the first cut (see Figure 6.1). In
     America this is known as a 'jump cut'.
(4) If however the branch which is to be removed is very heavy
     or very long, it should be removed either in several lengths
     beginning at the small or light end of the branch to be
     removed, or its weight should be supported by roping (see
     Figure 6.2).
(5) In making the third cut, as shown in Figure 6.1, Dr Shigo
     emphasizes that care should be taken to avoid cutting
     through the branch collar. On no account should cuts be
     made flush with the main trunk.
(6) When removing a secondary leader or a vertically growing
     branch, the cut should be finished with a slanting surface
     so as to throw off rain-water and other moisture.

(7) The removal of large branches from a tree can cause a check in growth, and where several branches have to be removed, the work should be spread over a period of time. This should minimize the effect.

(ii) *Lopping and topping*
(1) The term 'lopping and topping' is one which is more commonly used in forestry than in arboriculture.
(2) 'Lopping' is a term which refers to the removal of large side branches, while 'topping' is the removal of the crown or head of the tree.
(3) In forestry, the term 'lop and top' is used to indicate the branches which are removed when a large tree is being trimmed out after felling. It is especially applicable to large open-grown hardwoods which produce wide heavy crowns.
(4) In arboriculture the term 'lopping and topping' is not infrequently regarded as indicative of a crude form of heavy-handed pruning.

(iii) *Roping and slinging*
(1) In some cases the branch must be roped to prevent it damaging the tree itself, nearby buildings or other trees, in the course of an uncontrolled fall.
(2) The usual method is to pass a rope over a fork and attach the end to the branch which is to be removed. One end of the rope should be attached to the branch, slightly off the point of balance so that the heavy end tips towards the ground. The other end should be passed around the protected bole of an adjoining tree or other suitable object so as to control the run of the rope (see Figure 6.2).
(3) With very heavy branches two ropes may be necessary, and in any case the use of a lead or guide rope is very advisable in all but the simplest operations.
(4) The weight of the branch depends on several factors. These include its length, its diameter, the species of tree and the moisture content of the timber. The higher the moisture content the heavier is the timber and consequently the weight of growing timber is much more than that which has been felled and seasoned. Some hardwoods when green contain the same weight of water as they do wood.
(5) The following knots are generally used in tree surgery and timber work:

Clove hitch	Bowline
Timber hitch	Sheet bend
Carter's hitch	Prussik.

Excellent drawings of these and other knots will be found in *Tree Surgery* by P. H. Bridgeman (1976).

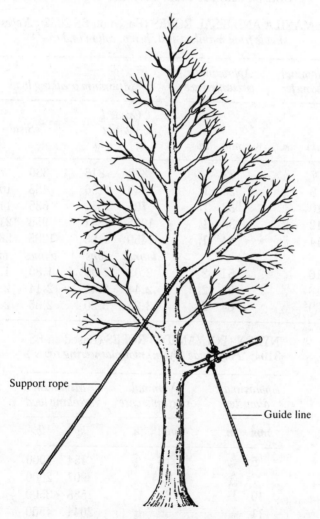

Support rope

Guide line

Figure 6.2 Slinging heavy limbs

(6) The safe working load, that is to say, the maximum load which should be put on a rope, should not be more than one-sixth of the breaking load. Thus with an 8 mm ($\frac{5}{16}$ in.) diameter grade 1 manila rope, the safe working load would be 90 kg (one-sixth of 543) or 200 lb. (one-sixth of 1196).

(7) The following two tables are based on figures contained in British Standards 2052: 1977 and 3104: 1970 (1987).

MANILA AND SISAL ROPES (Based on BS 2052: '*Ropes made from manila, sisal, hemp, cotton and coir*')

Nominal diameter		Approximate circumference		Minimum breaking load			
				Grade 1 manila		Sisal	
mm	in.	mm	in.	kg	lb.	kg	lb.
7	$\frac{9}{32}$	22	$\frac{7}{8}$	370	816	330	728
8	$\frac{5}{16}$	25	1	543	1196	483	1064
10	$\frac{3}{8}$	32	$1\frac{1}{4}$	705	1554	635	1400
12	$\frac{15}{32}$	38	$1\frac{1}{2}$	1065	2347	955	2105
14	$\frac{9}{16}$	44	$1\frac{3}{4}$	1450	3196	1285	2832
				tonnes	tons	tonnes	tons
16	$\frac{5}{8}$	51	2	2.03	2.00	1.80	1.77
18	$\frac{23}{32}$	57	$2\frac{1}{4}$	2.44	2.40	2.14	2.11
20	$\frac{25}{32}$	64	$2\frac{1}{2}$	3.25	3.20	2.85	2.80

NYLON (POLYAMIDE) ROPES (Based on BS 3104: '*Polyamide (nylon) mountaineering ropes*')

Approximate diameter		Nominal circumference		Minimum breaking load	
mm	in.	mm	in.	kg	lb.
5	$\frac{5}{32}$	16	$\frac{5}{8}$	454	1000
7	$\frac{9}{32}$	22	$\frac{7}{8}$	907	2000
10	$\frac{3}{8}$	31	$1\frac{1}{4}$	1588	3500
11	$\frac{7}{16}$	35	$1\frac{3}{8}$	2041	4500

(e) *Crown reduction*

(i) The reduction of the crown of a tree may be carried out either by lifting the crown or by lowering it.

(ii) Lifting the crown: this consists of either removing some of the lowest branches of the crown or removing parts of such branches, as shown in Figure 6.3.

(iii) Lowering the crown: this is sometimes known as dropcrotching and comprises the cutting back and shortening of the branches which form the perimeter of the crown. This is shown in Figure 6.4.

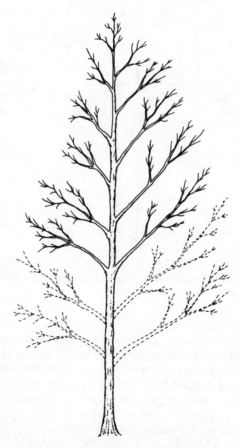

Figure 6.3 Lifting the crown (parts removed shown by dotted lines)

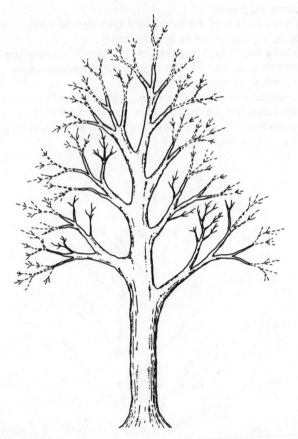

Figure 6.4 Lowering the crown or dropcrotching (parts removed shown by dotted lines)

(f) Pruning roadside trees

(i) Roadside trees, especially those which have been planted close to the adjoining carriageway, require special care and attention as regards pruning. In urban areas, street trees can affect the amount of natural light reaching nearby houses, can create problems regarding the care and maintenance of adjoining houses and can, at certain times of the year, reduce the effectiveness of artificial street lighting.

(ii) In the past the choice of trees for roadside planting has not always been of the best. Species have been planted which have become too large for the position for which they have been selected. In other cases, trees with a large spreading crown have been grown in situations which are quite unsuitable. In such cases the problem of pruning is very much accentuated.

(iii) Although the long-term answer to these problems is to choose a more suitable type of tree and to plant it in a more appropriate position, the fact remains that there are many existing trees which have to be dealt with in the meantime.

(iv) Trees with spreading crowns which have been planted close to the carriageway will frequently need severe pruning if damage to passing vehicles is to be avoided. Unfortunately this action, when carried out annually, can produce a very restricted head of deformed stumps.

(v) When planted too close to buildings trees can seriously obstruct natural light. The effect of this can be reduced by extensive pruning which may unfortunately detract from the appearance of the tree.

(g) Forked-growth

(i) The forking of the main stem of a tree can, in course of time, prove to be the source of both danger and expense. It is therefore essential for such form of growth to be dealt with as early as possible.

(ii) Dr Shigo advises that if water pockets occur in the fork of a tree, drainage holes should not be bored as these can cause decay.

(iii) Where a tree has developed a large fork which is considered to be dangerous, owing to the risk of it splitting apart, two alternative courses are open:

   (1) To remove one half of the fork. This can result in spoiling the appearance of the tree and producing an unbalanced crown.

   (2) To bolt and brace the two forked members so as to minimize the risk of splitting.

(h) Damage to roots

(i) Where roots have been broken or damaged, the broken ends should be trimmed off where possible.

(ii) If a large proportion of the roots have been severed, the stability of the tree may be affected and it may have to be removed.

2 *Treatment of wounds*

(a) *How wounds occur*
Wounds may be caused in the following ways:
- (i) By the removal of a branch during pruning.
- (ii) As a result of a branch breaking due to wind, snow, frozen rain, etc.
- (iii) Due to the fall of a large limb which may injure the main stem of the tree or other branches.
- (iv) By lightning. In such cases damage is usually caused to the bole of the tree through part of the bark being stripped off, although damage can be much more extensive.
- (v) Due to the death of a branch which subsequently falls and leaves a dead or rotting stump.
- (vi) As a result of damage by animals such as grey squirrels, rabbits or domestic animals.
- (vii) By vandalism.
- (viii) Due to machinery used in road work or for excavating trenches.
- (ix) By vehicles where trees overhang the highway.

(b) *The new approach to treatment*
- (i) Reference has already been made to the work of Dr Shigo, and detailed information will be found in the publications which are mentioned at the beginning of this chapter. However, it is possible to include only a brief summary of the main points in this section.
- (ii) Trees form protective barriers or boundaries around wounds and areas of decay which limit, but do not always halt, the spread of an infected area.
- (iii) It is most important that these barriers should not be broken when carrying out any work on the tree.
- (iv) Wounds and cut surfaces should no longer be treated with wound dressings or paint.
- (v) The removal of branches, flush with the main stem, must be avoided at all times so that the branch collar remains intact.
- (vi) If the bark surrounding a wound has been injured, any damaged pieces should be removed.
- (vii) As already noted, the boring of drainage holes or the insertion of drainage tubes should no longer be carried out.

3 *Bracing and bolting*

(*a*) *The need for bracing*
 (i) Bracing is a term used to describe the strengthening and supporting of a tree by means of cables and rods.
 (ii) Bracing is necessary under the following circumstances:
   (1) To provide support to large branches which might otherwise break on account of their weight.
   (2) To prevent or arrest the splitting of forked branches.
   (3) To brace the crown of a tree, i.e. to strengthen the crown of a large tree by means of braces between the main stem and its branches.
   (4) To reduce the effect of wind on the branches of a large tree.

(*b*) *Methods of bracing*
 (i) There are basically two methods of bracing:
     Cable or flexible bracing
     Rod or inflexible bracing.
 (ii) In addition there is the provision of ground support with props, which is sometimes known as vertical bracing.

(*c*) *Cable bracing*
 (i) *Principles*
     In cable bracing, support is given by means of wire ropes or cables, which provide a means of flexible support which allows for a certain amount of movement of the tree on account of wind.

 (ii) *Method of fixing*
   (1) In the first place it is necessary to decide where to affix the brace. It is generally considered that this should be attached two-thirds of the way up the branch, measuring from the main trunk. However, this must be regarded simply as a guide, and the actual position for the brace must be decided from inspection on the spot.
   (2) Having decided the position, it must be borne in mind that when fixed, the two eye bolts and the cable joining them must be in a straight line (see Figure 6.5).
   (3) The cable may be attached to the branches concerned by means of either eye bolts or screw-eyes, which are also known as screw hooks. Under no circumstances should a cable be wrapped around a limb nor should iron collars or bands be used.

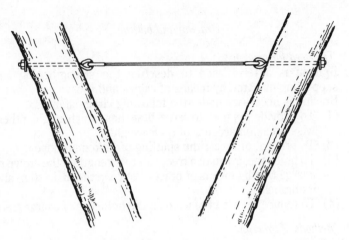

Figure 6.5 Cable bracing (Bolts and cable must be in a straight line.)

(4) Eye bolts, together with their nuts and washers, which
    should be oval or diamond-shaped, should be of galvanized
    steel and will vary in size according to the strain which they
    will have to stand.
(5) British Standard 3998: 1966 recommends a cable of
    galvanized steel stay wire (31-ton quality) which complies
    with British Standards 182 and 183 '*Galvanised iron and
    steel wire*' as follows:

    Branches up to 6 in. diameter
      (at point of attachment)
                          = British Standard Ref. No. 7/14 cable
    Branches up to 10 in. diameter
                          =   „   „   „   „ 3/8   „
    For larger branches
                          =   „   „   „   „ 7/8   „

(6) Eye-bolts are inserted into holes which have been drilled
    through the branch at the required angle. The washer
    which is placed in front of the retaining nut is countersunk
    into the branch.
(7) Screw-eyes, which are sometimes used instead of eye
    bolts, are screw-threaded eyes. A hole is first bored into
    the branch with an augur, about 2 mm ($\frac{1}{16}$ in.) smaller
    than the diameter of the eye, which is then screwed in.

Although doubt is sometimes cast on the holding power of a screw-eye, there is little risk of them pulling out if they are correctly fixed. Furthermore, as the tree grows they become more firmly embedded. The eye should be screwed into the limb to a depth equal to a half or two-thirds of the limb's diameter. To fix an eye bolt, a hole is bored through the branch and this may cause a slight weakening. With a screw-eye this is avoided.

(iii) *Attachment of the cable to limbs*

    (1) Cables should be spliced around a thimble before attaching them to eye bolts or screw-eyes (see Figure 6.6). This

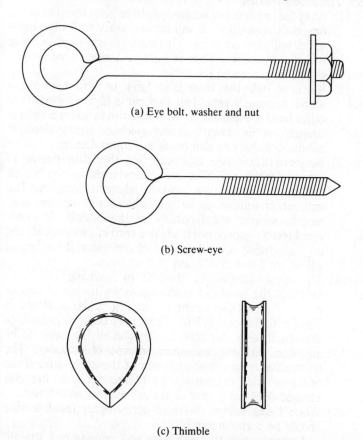

(a) Eye bolt, washer and nut

(b) Screw-eye

(c) Thimble

Figure 6.6 Eye bolt, screw-eye and thimble

protects the cable at the point where it passes through the eye. Detailed information on thimbles may be obtained from British Standard 464: '*Thimbles for wire ropes*'.

(2) An alternative to splicing is provided by galvanized steel wrap guy links manufactured by Preformed Line Products (Great Britain) Ltd. These are designed to facilitate the joining of the end of a cable after passing it round a pole or other object without using 'U' bolts. Full details can be obtained from the makers, whose address is East Portway, Andover, Hants SP10 3LH.

(iv) *Tensioning the cable*

(1) After the eye bolts or screw-eyes have been inserted in the branches concerned, it will be necessary to fix the cable which will connect them. There are several ways of doing this, but in the first place it must be decided what tension should be placed on the cable.

(2) If it is so tight that there is no 'give' or movement of the limbs, damage is very likely to occur in high winds. On the other hand if the cable is too slack, it may exert a sudden 'snatch' on the branch during gales or strong gusts of winds, and this can also result in serious damage.

(3) Between these two extremes lies the ideal degree of tension, which will allow some movement by the branch and at the same time provide adequate support. The amount of tension to be applied depends on the size, length, weight and flexibility of the branch. It needs considerable experience to make a correct assessment, and no opportunity should be lost of observing the effect of snow, wind and even heavy rain on trees.

(4) The most satisfactory method of securing the cable between the points of anchorage is by means of two or more 'U' clips (also known as 'bulldog' grips) as shown in Figures 6.7(a) and 6.7(b). The use of these clips enables the length and therefore the tension of the cable to be adjusted with the minimum amount of difficulty. The alternative is to splice each end of the cable around the thimbles after estimating the correct tension, but this method does not permit of any subsequent adjustment.

(5) When fixing cables, bolts or screw-eyes, the following should be borne in mind:

Always ensure that eye bolts and cable form a straight line when fixed.

When fixing cables, never allow them to touch any branch other than those to which they are fixed.

Good cabling is seldom seen if it is properly done.

(v) *Bracing the crown*

   (1) In addition to supporting and securing individual branches, cabling can be used to stabilize and safeguard the whole crown of a tree, and this is known as bracing the crown.

   (2) In crown bracing, cables are used to build up a pattern of support. This may take the form of lateral cabling between three or more main branches or stems, by means of which the crown, although strengthened, is allowed to move freely.

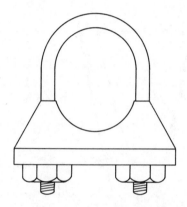

Figure 6.7(a) 'U' clip or 'bulldog' grip

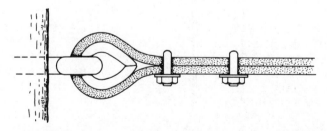

Figure 6.7(b) 'U' clips securing a cable passed through an eye bolt around a thimble

(*d*) *Rod bracing*
  (i) Unlike cabling, this form of bracing is not flexible and its use is
      limited to strengthening weak or split forks or securing two
      touching limbs which would othrwise rub and injure each other.
 (ii) In some cases a fork may be made additionally secure by means
      of a cable brace fixed some distance above the rod brace. In
      such cases the cable must be kept in full tension so as to
      prevent any movement which might put undue strain on the
      rods or allow the fork to split (see Figure 6.8).
(iii) In the case of two limbs which are touching and consequently
      rub and chafe when moved by the wind, a rod brace can be used
      to stabilize the branches and prevent movement. If the bark on

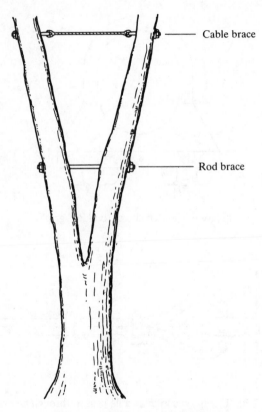

Figure 6.8 Rod and cable bracing

the adjacent limbs is pared off before bolting, grafting may result, which will strengthen the union (see Figure 6.9).

(iv) Care should be taken not to use too large a bolt or splitting may result. A. D. C. Le Sueur recommended a $\frac{1}{4}$ in. (6 mm) diameter bolt for a limb of 4 in. (10 cm) diameter and $\frac{5}{8}$ in. diameter (17 mm) for a 12 in. (30 cm) diameter limb.

(e) Ground supports

(i) Although ground supports or props cannot be described as bracing in the proper sense of the term, this method of support has been included in this section for convenience.

(ii) Ground supports are used to support branches of large trees which are growing a few feet above and roughly parallel to the ground. They are not ideal and it may well be preferable to provide cable brace support from above. However, ground supports can be useful in certain circumstances, if only as a temporary expedient.

(iii) Props should be placed so as to give the maximum amount of support, and with long heavy branches it may be necessary to

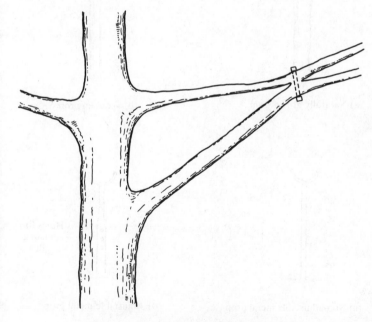

Figure 6.9 Stabilizing touching branches

use two or sometimes more in order to provide the necessary amount of underpinning.

(iv) Supports may be made of timber or metal. Timber props should be impregnated under pressure with a suitable preservative such as 'Celcure'. Metal props should be treated to reduce the risk of rusting.

(v) The actual point of support may be provided by provision of a 'U' or 'Y'-shaped prop. In the case of wooden supports these are frequently cut out of a naturally formed limb, as in Figure 6.10(a). If these are not available, supports can be built from sawn timber. Metal supports require a purpose-made 'U', which can be made adjustable if the lower portion telescopes into the upper part, a pin being provided to hold it in position as in Figure 6.10(d).

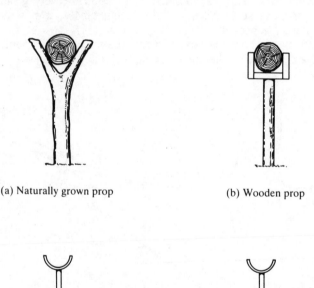

(a) Naturally grown prop   (b) Wooden prop

(c) Non-adjustable metal prop   (d) Adjustable metal prop

Figure 6.10 Ground supports

4 *Tools and equipment*

A list of some of the tools and equipment which are used in tree surgery are given below, but a full description of these and other items will be found in *Tree Surgery* by P. H. Bridgeman. Very useful information can also be obtained from the catalogues of firms which specialise in arboricultural equipment, such as Messrs Stanton Hope and Messrs Broadleaf Brown.

(*a*) *Saws*
  (i) *Power saws*
    Several makes of small light power saws ('chain saws') are now available which are particularly useful for arboricultural work.

  (ii) *Hand saws*
    There are a number of patterns which vary from single-handed pruning saws to bowsaws, the choice largely depending on working space and operational requirements.

  (iii) *Pole saws*
    Two main types of pole handles are fitted to these saws:
      Single-length wooden handles
      Aluminium or fibreglass handles which can be extended either by screw-jointed lengths of about 1.5 m (5 ft) long or by a telescopic arrangement.

(*b*) *Axes and related tools*
  Axes
  Hatchets
  Billhooks
  Slashers

(*c*) *Pruning tools other than saws*
  Short-handled secateurs also known as loppers
  Pole-handled pruners operated by a lever at the butt and working through a connecting rod or by a rope

(*d*) *Augurs and drills*
These can either be manual or power-driven, the latter being operated through a chain saw unit or by electricity, but this should not be used in wet weather.

(*e*) *Scrapers*
  Scrapers
  Chisels
  Excavating tools

(f) *Other items*
    Ropes
        Manila
        Sisal
        Nylon
        Wire
    Bracing rods
    Ladders
    Road signs: 'Tree cutting'
    Eye bolts
    Screw-eyes
    Thimbles
    Washers
    'Bulldog' grips ('U' clips)

(g) *Personal equipment*
    Safety belts and harness (see British Standard 1397)
    Climbing items including karabiners and climbing irons
    Safety helmets
    Visors and goggles
    Clothing: jackets and trousers
    Footwear
    Gloves
    First aid kit

5 *Fertilizing large trees*

(a) *The purpose of fertilizing*
(i) When trees become advanced in age their general health and well-being sometimes begins to fail. This may be due to various causes including:
    Old age
    Ill health
    A rise or fall in the water table
    Die-back of part of the root system
    Soil deterioration and a reduction in the supply of plant food.
(ii) Some of these conditions can be improved by the application of fertilizers.

(b) *Types of fertilizers*
(i) Fertilizers for large trees can be divided broadly into two classes:

Firstly, organic manures such as leaf mould, farmyard manure and so on

Secondly, artificial manures.

(ii) In the case of artificial manures, these should provide a proportion of nitrogen, phosphate and potash. A. D. C. Le Sueur in *The Care and Repair of Ornamental Trees* recommended the following complete manure:

Super phosphate (40%)	203 kg	(4 cwt)
Nitrosoda	330 "	(6½ ")
Sulphate of potash	127 "	(2½ ")
Dried blood	356 "	(7 ")
	1016 "	(20 ")

British Standard 3998: 1966 gives the following example:

Super phosphate (18% P_2O_5)	5	parts by volume
Ammonium sulphate (20% N)	5	" " "
Potassium sulphate (48% K_2O)	1.5	" " "
	11.5	

mixed with 5 parts of peat or 10 parts of sand to one part of the above mixed fertilizers.

There are several proprietary brands of compound fertilizers which are specially produced for trees and shrubs of which the following analysis is an example.

Nitrogen (N)	10.00%
Phosphate (P_2O_5)	7.50%
Potash (K_2O)	10.20%
Magnesium oxide	3.00%
Iron	0.25%
	30.95%

with traces of manganese, copper, boron and molybdenum.

(c) *Methods of application*

(i) When applying fertilizers to large trees, the most satisfactory results will be obtained if the material is placed as close to the roots as conditions will allow. Fertilizers can be applied as described in the following paragraphs.

(ii) By application to the surface of the ground surrounding the tree: this method is sometimes used when applying organic manures such as compost, leaf mould or farmyard manure. The disadvantages are that it is unsightly and does not readily come into contact with the roots. On the other hand it is a cheap and easy way of application.

(iii) By making holes in the ground around the tree: the holes should be 3–4 cm ($1\frac{1}{4}$–$1\frac{1}{2}$ in.) diameter and are usually made with a crowbar, but small power-driven augers are sometimes used. The depth of the holes should be 46–61 cm (18–24 in.), the deeper holes being used for species with a deeper root system. The fertilizer is placed into the holes, which are then topped up with soil.

In order to obtain the best results holes should be made in a belt approximately 3 m (10 ft) wide, the centre of which corresponds to the outline of the maximum circumference of the crown. The holes are spaced at intervals of 46–61 cm (18–24 in.) (see Figure 6.11).

Note. Another version of the above whereby small holes are made with a fork and the fertilizer scattered on the ground and then swept into the holes is not to be recommended. Much of the fertilizer remains on the surface, and it is doubtful if much of it reaches the roots in any strength.

(iv) By removing the top soil to a depth of about 15 cm (6 in.) and applying an organic manure such as farmyard manure or leaf mould. This method is the most expensive, and it is sometimes objected to on the grounds that when carried out it causes disfigurement. The dimensions of the belt are the same as those shown in Figure 6.11. After treatment, the area should be well watered.

(*d*) *Rates of application*
  (i) A. D. C. Le Sueur gives the following rule for estimating the amount of fertilizer to be applied to a tree:
  (a) *Middle-aged and old trees*
    (1) 3 lb. per inch diameter of the trunk at breast height, that is at 5 ft. Thus for a 20-in. diameter tree, this would be 3 × 20 or 60 lb.
    (2) The metricated equivalent of this would be 1.36 kg per 2.5 cm of diameter at 1.5 m. Thus for a 51 cm diameter tree the calculation would be:

$$\frac{1.36 \times 51}{2.5} \text{ or } 27 \text{ kg.}$$

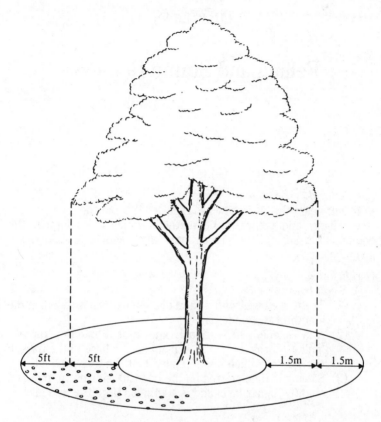

Figure 6.11 Application of fertilizers (insertion holes shown on only one quarter of area to be fertilized)

(b) *Trees under 15 cm (6 in.) diameter*
These should receive half the amounts as calculated above.
(ii) The application of 1.36 kg (3 lb.) which is given in the above example, can be varied from 0.45 to 2.27 (1 to 5 lb.) according to the condition of the tree, the soil and so on.

(*e*) *Time of application*
(i) Artificial manures should be applied in early spring so that the tree can take advantage of them when growth begins.
(ii) Organic manures such as compost and farmyard manure should be applied in late October or November. These should provide a covering of 10–15 cm (4–6 in.) deep.

CHAPTER VII

Felling and Stump Removal

1 *Felling*

(*a*) *General*

In recent years the chain saw has replaced the axe and cross-cut for tree felling, and some of the methods which are adopted, the equipment used and the precautions which should be taken, are given below.

(*b*) *Felling*

(i) *Small trees*

(1) These are trees with a basal diameter which is less than the length of the guide bar.

(2) A 'V'-shaped cut, variously known as a 'mouth', 'throat', 'gullet', 'beak' or 'sink', is made at the base of the tree on the side on which it is intended that the tree shall fall.

(3) Sawing is then started on the opposite side of the tree at a slightly higher level and is continued until the tree falls.

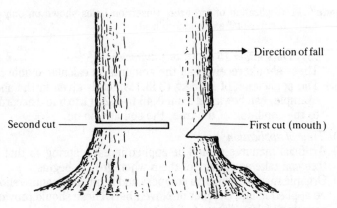

Figure 7.1 Felling a small tree

(ii) *Medium-sized trees*
   (1) These have a basal diameter equal to $1-1\frac{1}{2}$ times the length of the guide bar.
   (2) After the mouth has been made, the main cutting operation is begun by using the end of the saw and then working round in a clockwise direction, as if following the spokes of a wheel.
   (3) The sequence of action is shown by the numbers and arrows in Figure 7.2(a), a 'hinge' being left next to the mouth.

(iii) *Large trees*
   (1) Large trees are those which have a basal diameter equal to $1\frac{1}{2}-2\frac{1}{2}$ times the guide bar length.
   (2) After making the 'mouth' ('1' in Figure 7.2(b)), it is enlarged by cutting a further area into the centre of the tree ('2' in Figure 7.2(b)).
   (3) The saw is then withdrawn and, starting at one side, cutting is carried out in a clockwise direction. The sequence is shown in the diagram by numbers and arrows.

(*c*) This section provides only a brief summary of the procedures which should be followed when felling trees with a chain saw. For further detailed information, reference should be made to Forestry Safety Council Guide No. 11, *Felling by Chain Saw*, and Forestry Commission Leaflet No. 75, *Harvesting of Windthrown Trees*.

(*d*) *Equipment*
The equipment which should be used in order to ensure safe and efficient work when using a chain saw, may be divided into two categories:

(i) *Personal equipment*
   This should include:

Safety helmet	Safety gloves
Visor for eye protection	Safety boots
Ear protectors	Snag-proof clothing.
First aid kit	

   For further information regarding safety see Chapter VIII, *Safety in Arboriculture*.

(ii) *Technical equipment*
   In addition to the saw, which should be provided with a guard, the following items are amongst those which are normally required:

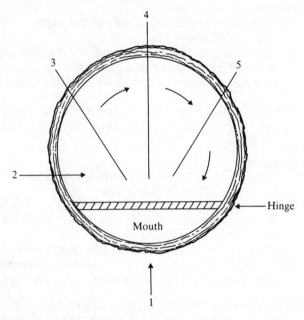

Figure 7.2(a) Felling medium-sized trees by power saw

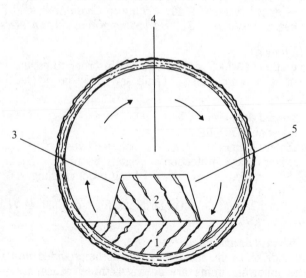

Figure 7.2(b) Felling large trees by power saw

Wedges	Cant-hook
Hammer	Breaking bar
Felling tongs	Fuel and oil cans.

A light hand-operated winch, such as the Tirfor, can also be of considerable assistance when dealing with trees which are hung-up.

(e) Chain saw maintenance

(i) Chain saws are carefully designed and accurately constructed machines and it is essential to maintain them properly if they are to start easily, cut efficiently and operate reliably.

(ii) Full instructions are issued by the manufacturers of the numerous chain saws which are on the market, and these instructions should be carefully studied.

(iii) An excellent manual which deals exclusively with the maintenance of chain saws has been published by the Oregan Saw Chain Division of Messrs Omark UK Ltd. Single copies can be obtained free of charge from this firm at 6 Station Drive, Bredon, Tewkesbury, Glos. GL20 7HQ.

(f) Felling precautions

(i) *Trees growing on slopes*

Trees on steep slopes should not be felled uphill as there is a considerable risk that the butt may 'jump back' and injure the faller. On moderate gradients trees may be felled 'across the slope'.

(ii) *Hung-up trees*

These are trees which, as they are being felled, fall against a standing tree and become lodged in it. They can be released with a cant-hook, lever, felling tongs or winch. Attempts should never be made to dislodge a hung-up tree by jumping on the bole, by felling the tree in which it has become jammed, by cutting off part of the butt or by felling another tree across it. Further information will be found in Forestry Safety Council Guide No. 14, *Takedown of Hung-up Trees*.

(iii) *Felling large trees*

Special precautions should be taken when dealing with large trees, particularly heavily crowned broadleaves standing on open sites. Felling should not be undertaken in a high wind as this may cause a change in the direction of fall. Where overhead electric or telephone wires are close to a tree, the authority concerned should be contacted beforehand. Large trees should always be examined for signs of butt rot before beginning work.

See also Forestry Safety Council Guide No. 17, *Felling Large Hardwoods*.

(iv) *Felling across obstacles*
Trees should not be felled across banks, walls or other trees since there is considerable risk that they will be broken or damaged by such action.

(g) *Other chain saw operations*
(i) In addition to felling trees, chain saws can be used for many other purposes, and these include snedding, cross-cutting logs and poles and clearing windblow.

(ii) Each of these operations requires a specialized technique, and detailed instructions will be found in the following guides issued by the Forestry Safety Council:
No. 12, *Chainsaw Snedding*.
No. 13, *Cross-cutting and Stacking*.
No. 15, *Chainsaw Clearance of Windblow*.

2 *Stump removal*

Stumps may be dealt with by physically removing them, by reducing them in size so that they can be covered with soil and are no longer apparent, or by treating them with chemicals so as to kill them and indirectly make them easier to remove. Stump removal may be considered under the following four headings:

(a) *Grubbing and extraction*
(i) *Removal by hand*
(1) It will often be necessary to dig around the largest side roots to enable them to be cut through and so facilitate the removal of the stump.
(2) Extraction is usually carried out with a hand-operated winch such as the Trewhella Monkey Winch or the Tirfor.
(3) In some cases, however, levers and jacks may have to be used where the operating conditions preclude the use of a winch or where a suitable anchor cannot be found or provided.

(ii) *Removal by machinery*
The removal of stumps with the use of mechanical equipment can be considered under three sections:
(1) *Preparatory work*
In the case of large stumps it is often advisable to carry out some preparatory work. This may consist of removing the

soil around the stump with a mechanical excavator and cutting the largest side roots. Where very large stumps have to be removed it may be necessary to use explosives in order to split the stump up into pieces which can be conveniently handled. The use of explosives is referred to later on in this section.

(2) *Mechanical excavating*

Under suitable conditions a mechanical excavator can remove stumps quickly and cheaply but a certain amount of damage to the surrounding ground is bound to occur. However this can usually be made good by in-filling and turfing over or seeding down, but it is generally inadvisable to carry out this method of stump extraction when conditions are very wet and the ground waterlogged.

(3) *Winching out*

Stumps can also be removed by using a tractor-mounted power winch, although a certain amount of preparatory work may be necessary when dealing with the bigger stumps. It is sometimes advisable to use a limited amount of explosives on very large stumps, not only to assist in their extraction, but also in their disposal after they have been pulled out.

(*b*) *Stump cutting*

(i) Stump cutting or grinding is carried out with a machine which consists basically of a power-driven vertical cutting wheel to which specially hardened 'teeth' are fitted. As this revolves, it moves gradually across the stump and in doing so, reduces it to chips.

(ii) A machine can reduce a stump 25–30 cm (10–12 in.) high to a depth of about 61 cm (24 in.) below ground level depending on the size of the machine used.

(iii) The larger machines are fitted with road wheels and can be towed behind a Land Rover or its equivalent, while the smaller models are mounted on two wheels and can be moved around the site by hand. A small trailer is used for transportation between different locations.

(iv) The advantages of this method of stump removal are that neither heavy excavating machinery nor explosives are necessary.

(*c*) *Explosives*

(i) Under suitable conditions, the use of explosives for stump removal has several advantages. Very large stumps can be removed while the amount of soil disturbance can be less than in the case of mechanical extraction, and at the same time the stump can be reduced to a manageable size for removal.

(ii) On the other hand explosives should not be used near houses, and considerable care must be taken as regards their use if they are to be used safely. The successful and safe use of explosives needs skill and experience, and it is best to obtain the services of a firm who specialize in this kind of work.

(iii) Holes are bored with a 5 cm (2 in.) diameter augur and 'cartridges' of gelignite are placed in them with a detonator. Charges are usually fired electrically, but if a burning fuse is used, care should be taken to ensure that only 'safety fuse' is utilized. This burns at the rate of 183 cm (6 ft) per minute and is white in colour. Before using it, a test length should be cut off and ignited to see that the burning time is correct.

(iv) The amount of gelignite used is 0.68 kg (1½ lb.) per 30 cm (12 in.) diameter of the stump, and after inserting the cartridge in the hole, the end should be filled with soil and gently firmed down. Before detonating, all persons in the immediate area should take cover.

(v) Before explosives can be used, a permit must be obtained from the Superintendent of Police, and if more than 4.5 kg (10 lb.) of gelignite is to be retained for use, a special magazine must be built.

Note. For further information on the subject of stump removal reference should be made to the following publication:

Removal of Tree Stumps – Forestry Commission Arboricultural Leaflet No. 7.

(*d*) *Chemicals*

(i) Although stumps can be killed by means of chemicals, there is no chemical which will quickly and effectively rot a stump. This means that even if a stump is killed by the application of a chemical, its removal will depend on the use of one of the methods which have been described earlier in this section. It is probable that a dead stump will offer less resistance to its removal than a live one with growing roots.

(ii) Hardwood stumps can be killed by the application of either 2, 4, 5-T or Ammonium sulphamate.

(iii) 2, 4, 5-T

(1) This chemical is very effective for killing stumps, and these may be sprayed or brushed with 2, 4, 5-T at a rate of 6.8–

9.1 litres per 455 litres (1½–2 gallons per 100 gallons) of diesel oil.

(2) Spraying can be carried out at any time so long as the stumps are not wet with rain. The rate of application varies, but a 51 cm (20 in.) diameter stump will require about 0.6 litres (1 pint) of spray. It is important that the application should saturate the sapwood and bark.

(iv) *Ammonium sulphamate*

(1) This can be applied in a solution of 1.8 kg (4 lb.) of ammonium sulphamate to one gallon of water by spraying or brushing. As in the case of 2, 4, 5-T, the bark and sapwood should be thoroughly saturated as far as possible.

(2) Ammonium sulphamate can also be applied in the form of dry crystals to the surface of the stump. In such cases the rate of application should be 0.28 kg (½ oz.) per 2.5 cm (1 in.) of stump diameter. Care must be taken to confine the application to the stump concerned or damage may be caused to other trees or plants in the immediate vicinity.

(v) It will be appreciated that research is being carried out continuously, and improved materials will undoubtedly be put on the market as they become available. The details and rates of application given above should be regarded only as a guide, and in all cases the directions issued by the manufacturers must receive priority and should be carefully studied before using the material.

(vi) Another material which is sometimes recommended for killing stumps is sodium chlorate, but this will cause the subject to which it is applied to become very inflammable. In dry weather spontaneous ignition may occur, and for this reason it is not altogether satisfactory.

Note. For further information regarding the use of chemicals on stumps reference should be made to:

Removal of Tree Stumps – Forestry Commission Arboriculture Leaflet No. 7.

Recommendations for Tree Work – British Standard 3998: 1966.

CHAPTER VIII

Safety in Arboriculture

1 *General*

(*a*) The felling and heavy pruning of trees has always been potentially dangerous work, but in recent years the opportunities for accidents in other operations have increased considerably.

(*b*) This is largely due to:

(i) The widespread use of machinery, notably the chain saw.

(ii) The introduction of new types of equipment for specialized operations such as stump cutting, brush and wood chipping, and log splitting.

(iii) The more general use of chemicals, especially herbicides.

(*c*) In arboriculture, there is an increased risk of accidents on account of the circumstances and localities in which work is carried out. Since this is frequently in urban areas, the presence of the public, traffic and buildings can all contribute to the problem.

2 *Safety precautions*

(*a*) Work which involves the climbing of trees can always be dangerous, and every step should be taken to reduce the risk of accidents.

(*b*) Set out below are some of the points which should be borne in mind when men or women are working in trees:

(i) Always take adequate safety precautions, whether climbing, pruning, felling or loading.

(ii) When climbing in trees, always wear a harness or safety belt and safety helmet, together with the necessary ropes.

(iii) Remember that the weather can have an important bearing on working conditions, and never climb trees in a high wind or when the branches are wet or are covered by frost or snow.

(iv) Never fell a tree in a high wind, and trees growing on a slope should always be felled downhill.

(v) When climbing, move slowly and with care. Do not jump on branches to see if they are sound, and remember that the branches of some species tend to be weak or brittle. Test handholds before using them.

(vi) When climbing, wear a boiler suit or overalls without a belt, loops or anything that can catch on snags. Never wear a scarf or sweat rag around the neck. Boots can be a problem: avoid those with soles which can slip or slide on branches. Rubber soles with a heavy tread are considered to be the best.

(vii) Avoid any 'slack' forming in the safety rope. If this happens and a fall occurs, the resulting jerk when the slack is taken up can have very serious consequences.

(viii) If ladders are used, they must be set on a firm base and at some point should be lashed to the tree. A slipping ladder is very dangerous.

(ix) Whenever a man or woman is working in a tree, there should always be another worker on the ground in case of accidents.

(x) No worker should be allowed to operate in a tree until he/she has been properly trained and has had adequate supervised experience of the kind of work to be undertaken.

(xi) Remember that even an experienced worker can make an error of judgement, and it is better to be safe than sorry. The higher they climb, the harder they fall.

(c) Further information on various aspects of safety, including equipment, clothing, tree climbing and the public, will be found in *Tree Surgery* by P. H. Bridgeman, 1976.

3 Safety guides

(a) When the Forestry Safety Council was set up by the Forestry Commission in April 1974 to promote safety in forestry, its terms of reference included the production of Forest Industry Safety Guides.

(b) In 1989, 29 guides were available which covered a number of operations as well as the use of machinery.

(c) Although these guides are intended for forestry, the following ones would be of value in arboriculture:

1. *Clearing Saw*
10. *The Chain Saw*
11. *Felling by Chain Saw*
12. *Chain Saw Snedding*
13. *Cross-Cutting and Stacking*
14. *Takedown of Hung-up Trees*
15. *Chain Saw Clearance of Windblow*
17. *Felling Large Hardwoods*

18. *Tree Climbing and Pruning*
26. *Use of Tractors with Winches in Directional Felling and Take-down*
30. *Mobile Saw Bench*
33. *Hand Held Power Posthole Borer*
34. *First Aid*
N. *Noise and Hearing Conservation.*

Copies of these can be obtained, free of charge, from the Secretary, Forestry Safety Council, Forestry Commission, 231 Corstorphine Road, Edinburgh EH12 7AT.

4 *Arboricultural Safety Council*

(*a*) In 1988, on the initiative of the Arboricultural Association, the Arboricultural Safety Council was set up.

(*b*) It is composed of representatives of various bodies, which include the following:

Arboricultural Association
Association of Professional Foresters
Forestry Commission
Forestry Training Council
Health and Safety Executive
Institute of Chartered Foresters
Royal Forestry Society of England, Wales and Northern Ireland.

(*c*) The objectives of the Council may be summarized as follows:
(i) To promote safety in arboriculture.
(ii) To cooperate with the Health and Safety Commission.
(iii) To produce Arboricultural Safety Guides.
(iv) To maintain a close connection with the Forestry Safety Council, the Forestry Training Council and other appropriate bodies.
(v) To provide publicity on safety matters.

5 *Health and Safety at Work Act 1974*

(*a*) *General*
(i) This Act sets out the law relating to the health, safety and welfare of those who are employed or self-employed, and also members of the public who might consequently be affected.
(ii) On 1 October 1974, the Health and Safety Commission was set up under the Act. This body has the duty of developing health and safety, and drawing up new regulations when necessary.

(*b*) *Duties imposed by the Act*
 (i) Employers must, so far as is reasonably practical, safeguard their employees by:
   (1) providing safe plant and machinery;
   (2) adopting safe systems of work;
   (3) ensuring proper training, instruction and supervision.
 (ii) Employees and self-employed persons must take reasonable care of their own health and safety, and also that of the members of the public who may be affected by their work.
(iii) Manufacturers and suppliers must provide machinery and equipment which is safe to use.

(*c*) *Notices and inspectors*
 (i) Two types of notices may be served by inspectors appointed under the Act:
   (1) Improvement Notices which state what action should be taken.
   (2) Prohibition Notices which require a complete or partial stoppage of work, if action required by an Improvement Notice is not taken within a given time. The Notice can require an immediate stoppage if there is a serious safety hazard.
 (ii) Appeals against notices can be made to an Industrial Tribunal.

(*d*) *Codes of practice*
 (i) Regulations may be supplemented by codes of practice which can be used in criminal proceedings as evidence of the infringement of a statutory requirement.
 (ii) The Health and Safety Commission can prepare codes of practice and approve those which have been drawn up by other authorities.

6 *Safety committees and representatives*

Regulations relating to safety committees and safety representatives came into force on 1 October 1974 and the effect of these may be summarized as follows:

(*a*) *Safety representatives*
A recognized trade union can appoint safety representatives from amongst the employees of an employer who recognizes the union. Such representatives should have at least two years' experience in their particular work.

(*b*) *Safety committees*
(i) These can be set up where two or more employees ask their employer to do so.
(ii) The membership of the committee should be agreed by the management and the union members.
Further information will be found in the Health and Safety Council's booklet *Safety Representatives and Safety Committees*, which can be obtained from HM Stationery Office.

7 *Statutory Instruments concerning safety*

The following Statutory Instruments should be noted:

SI 1957/1386	The Agriculture (Power Take-off) Regulations 1957.
SI 1959/427	The Agriculture (Circular Saw) Regulations 1959.
SI 1959/1216	The Agriculture (Stationary Machinery) Regulations 1959.
SI 1962/1472	The Agriculture (Field Machinery) Regulations 1962.
SI 1971/694	The Road Vehicles (Lighting) Regulations 1971.
SI 1973/24	The Motor Vehicular (Construction and Use) Regulations 1973
SI 1981/917	Health and Safety (First Aid) Regulations 1981.
SI 1985/2023	The Reporting of Injuries, Disease and Dangerous Occurrences Regulations 1985.

CHAPTER IX

Street and Roadside Trees

1 *Points to consider*

The planting and tending of trees in streets and on roadsides are affected by special considerations which arise out of the unnatural circumstances in which such trees often grow. Some of the points which must be borne in mind are set out below.

(a) *The width of the street or road*
 (i) The wider the street the more room there is to plant trees and to allow them to grow naturally. On the continent of Europe there are many examples of wide streets which are lined with two or more rows of trees forming magnificent vistas and avenues or providing welcome shade on hot summer days. Although in this country there are a few examples of such wisdom, it is unfortunate that they are conspicuous by their absence.
(ii) The amount of space available controls the type of tree which can be planted. In the narrower streets columnar and fastigiate trees may have to be used and those which do not grow to any great size. In really narrow streets tree planting cannot be undertaken.

(b) *Proximity to buildings*
 (i) In towns, trees should never be planted in close proximity to buildings. If this is done the tree may tend to grow outwards into the street causing an obstruction to traffic, but more likely it will be cut back heavily so that its shape is spoilt and frequent pruning may subsequently become necessary. At the same time the amount of light which would otherwise reach the windows of the adjoining building may be obstructed.
(ii) If trees are planted close to a house, there is always a risk of the roots causing damage to the drains and foundations. This is especially so in the case of poplars, and this aspect is considered in Chapter XV.

(c) Street lighting
(i) Trees can affect the efficiency of street lighting, and this should be borne in mind when planting them.
(ii) Similarly when a new system of street lighting is being installed, the existence of any trees which might affect the scheme should be taken into account.

(d) Traffic requirements
(i) The operation of double-decker buses, large lorries and vans means that trees must either be planted well back from the roadway, or if they have already been planted, they must be pruned back and any obstructing branches removed.
(ii) On bends, corners and road junctions, trees which obstruct visibility must be pruned back, or where this does not produce the required result, they may have to be removed entirely.

(e) Public services
Public services such as water pipe lines, sewers, electricity, gas and telephones generally follow the course of the street or road in urban areas. When these require repair or maintenance it is often necessary to excavate the road, footpaths or verges in order to reach the necessary pipe or cable. In doing so, the roots of the adjoining trees can be seriously disturbed, and unless care is taken, this may produce fatal results.

(f) Physical shortcomings
(i) In addition to the points which have already been mentioned, trees in streets can suffer in other ways.
(ii) The soil in which they are planted can well have been of poor quality and possibly 'made' ground which has been tipped on the site, so that it is of very low fertility.
(iii) Owing to the presence of paved footpaths and tarred roads, little water reaches the roots of many street trees. When it does, it may well be partly contaminated.
(iv) The air of many large towns is polluted by diesel and petrol fumes which may adversely affect the health of the tree.

2 *The prerequisites of street and roadside trees*

When selecting trees for street and roadside planting, it is wise to ensure that they possess certain characteristics which, in themselves, make the tree suitable for their environment. These characteristics or prerequisites are summarized below.

(*a*) *Size*

 (i) Very large trees are seldom suitable for street planting unless the layout is sufficiently spacious to permit their use. In some of the wide boulevards which are to be seen on the Continent large trees are used with considerable success.

 (ii) On the whole, trees used in street planting must remain comparatively small when they reach their full size.

 (iii) For roadside planting in less densely built-up areas larger trees can be used because more space is generally available for tree planting. In rural areas the restriction on size ceases to exist in so far as space is concerned. Any restrictions will be due to other considerations such as risk of wind blow, visibility and so on.

(*b*) *Shape and form*

 (i) The shape and form of a tree is to some extent related to the question of size. It is possible that in certain circumstances a large fastigiate or columnar-shaped tree would be acceptable while a smaller tree with a wide spreading crown would not be.

 (ii) Generally speaking, there is little space available in the average street for trees with broad crowns. If such trees are planted they are, of necessity, often subjected to continual pruning which often produces clusters of unsightly swellings. Consequently there is much to be said for planting trees which have narrow crowns. At the same time one must bear in mind the optical effect which is produced by planting trees with a narrow pointed appearance.

 (iii) There are, however, several variations in the shape of narrow-crowned trees, and these are shown diagrammatically in Figure 9.1.

(*c*) *Appearance*

 (i) Although the height, breadth and shape of a tree all contribute to its appearance, in this section the word is used in respect of four specific features, namely flowers, fruit, foliage and bark.

 (ii) Flowering trees are planted very extensively in urban areas and during the flowering period they have much to commend them. Unfortunately the flowers remain on the trees only for a short time during the year, and when they are over the trees may have very much less to recommend them. Ideally the ability to produce handsome flowers should be combined with some other feature such as attractive foliage.

 (iii) In many ways attractive foliage is to be preferred to flowering ability, since in the case of deciduous trees, the leaves should

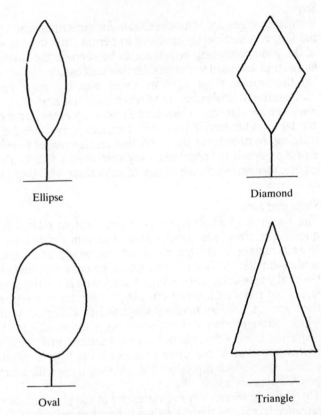

Figure 9.1 Diagrammatic variations in narrow-crowned trees

remain on the trees for at least six months of the year. Foliage can be variegated, coloured or display green on the upper surface and grey below, as for example in the European white lime (*Tilia tomentosa*). On the other hand it may be green throughout and rely on the size or shape of the leaves to emphasize its difference, as in the Indian Bean tree (*Catalpa bignonioides*). Autumn colour, although very striking, is short-lived and in some cases dependent on the season.

(iv) Fruit can add to the appearance of a tree during the season, but the production of fruit can sometimes be a disadvantage. For example, chestnuts or 'conkers' can result in damage being caused to the trees by children.

(v) An unusual bark can be very striking but in some cases it may encourage damage. The bark of some of the birches peels readily, and this can mean that it will be deliberately removed. The soft spongy bark of the Wellingtonia and California Redwood can also be subject to quite considerable damage, although it must be added that these species are unlikely to be used in street planting.

(d) Other considerations

In addition to the foregoing, trees selected for street or roadside planting should ideally be fast-growing in the early stages so as to discourage damage, able to withstand pruning, and sufficiently robust to overcome urban conditions.

3 Planting considerations

(a) The siting of trees

Street or roadside trees should *not* be planted:
  (i) Opposite the main front room windows of houses.
  (ii) Opposite garden gates.
  (iii) Near the entrances to existing garages or possible garage sites. Trees must not obstruct the visibility of drivers of vehicles leaving the garage.
  (iv) Near street lamps of the upright or standard design. Trees should be at least 6 m (20 ft) from a lamp post, except in the case of centrally hung and overhead lighting.
  (v) Too close to the curb, or damage may be caused to and by large vehicles such as double-decker buses.
  (vi) In a position which obscures the view of drivers of road vehicles whether at road junctions, on bends or elsewhere.
  (vii) Beneath or in close proximity to overhead telephone lines or electric cables.
  (viii) Near underground service lines such as gas and water mains, sewers, power and telephone cables. Excavation to repair or improve such services can cause serious damage to the roots of trees growing in the immediate vicinity.

It will be appreciated from the above that the choice of sites for street and roadside trees is, to say the least, restricted. However, it is better to site trees carefully so that they will be able to grow, rather than plant them in positions where they may be damaged, mutilated or even felled when they reach a worthwhile size.

(b) *Planting procedure*
(i) Street trees can be planted in two ways:
    (1) Regularly, at fixed predetermined distances.
    (2) Irregularly, at varying distances
(ii) From the previous section, dealing with the siting of trees, it will be clear that the planting of trees at fixed distances will be very difficult in some cases. This will largely depend on the amount of space available between the buildings and the edge of the highway. Generally speaking, the wider this is, the more scope there is for tree planting. A planting distance of about 18 m (60 ft) has frequently been adopted for street trees.
(iii) With irregular planting, the siting of trees is very much easier and the most suitable position can be chosen, irrespective of the distance to the next tree. Moreover any impression of regularity which may border on monotony is avoided. Irregular planting can create a much more natural appearance and at the same time reduce the number of trees which would be required for regular planting.

(c) *Planting method*
(i) The planting of trees is dealt with in Chapter II and reference should be made to that chapter for details of procedure.
(ii) Where trees are planted in pavements as opposed to roadside verges, as large a planting hole should be dug as is compatible with other conditions and requirements. One of these will be removal of existing paving stones prior to digging the hole. The size of hole will have to be a multiple of the number of stones removed, and this will depend on the size of the stones.
(iii) Steps should be taken to ensure that the tree is planted in a good fresh loam and this should be introduced into the planting hole, if necessary together with a little bone meal. However the cost of this may prove to be considerable if a large number of trees are to be planted. The depth of the hole should be 61–91 cm (24–36 in.).
(iv) After planting, it may be necessary to replace in part the paving stones which have been removed. This can be done by means of tree grilles which allow a certain amount of water to percolate on to the soil surrounding the tree, at the same time protecting the roots.

4 *Care and protection*

Staking, tying and after-treatment are dealt with in Sections 9 and 10 of Chapter II. Chapter IV covers in greater detail the matter of tree injuries and how to afford protection against such injuries. Reference should be made to the appropriate sections of these two chapters for information on care and protection.

5 *Planting on trunk roads and motorways*

(*a*) *General considerations*
  (i) If appropriate action is taken, a newly constructed road can be absorbed into the landscape and not allowed to form an ugly gash across the countryside.
 (ii) It is even possible for a road to contribute to the beauty of the rural scene, if it is well planned and steps are taken to blend it with the scenery. However, since engineering requirements and cost considerations are usually paramount, this happy state of affairs is less common than it could be. A well-designed road ought to be attractive to both the road user and the casual observer from afar.
(iii) In building a new road, attention should be paid to existing features, and advantage should be taken of them. Heavy-handed attempts at crude camouflage by unimaginative planting has little to be said for it.
 (iv) The construction of new highways or the improvement of existing roads will often produce areas of surplus land which fall within the boundary of the highway. Many of these provide admirable sites for tree planting which can greatly improve the appearance of the vicinity.

(*b*) *Visibility*
  (i) When planting trees beside, or adjacent to, a road, care must be taken to avoid any obstruction of the visibility for road users. Although this is especially the case at road junctions, it is equally important at corners and even on comparatively gentle bends.
 (ii) Trees must not be allowed to obstruct road signs or traffic lights.

(*c*) *Safety aspects*
  (i) In addition to visibility requirements, trees must be set well back from the edge of the highway. This will reduce the risk of

fatal accidents when vehicles skid, and may tend to lessen the amount of leaves which fall on the road surface, although the effect of windblown leaves will often neutralize this point.

(ii) The view has been expressed that rows of trees planted along the sides of long straight roads, tend to produce drowsiness and lack of alertness in drivers. How much of this is due simply to driving along a straight level road for mile after mile and how much is due to the presence of trees is a matter of opinion. The fact remains that driving along a motorway which has no sizeable trees along it can produce a sense of lethargy, especially on a hot day.

(iii) Trees should not be planted in such a position that when they have reached sizeable proportions, they would fall on to the motorway should they be blown down.

(d) *Choice of species*
  (i) Since motorways and trunk roads pass, for the greater part of their length, through the countryside, the species chosen for planting beside them should, as far as possible, be those species to be found in the countryside.

(ii) If, for some reason, this is not possible, care should be taken to plant 'country' trees and not 'town' trees such as ornamental flowering species, which look quite out of place in the British countryside.

(iii) Where it is intended to plant slow-growing hardwoods such as oak and beech, these may be mixed with a faster-growing species which will not only provide earlier intermediate effects, but will also act as a nurse to the slower-growing trees. In due course the nurses can be removed, leaving the hardwoods in possession of the site. Conifer nurses, particularly larch, can often provide the most satisfactory results in such cases.

(e) *Establishment*
  (i) In many cases trees of the size normally used in forestry can be planted. These vary from 23 to 46 cm (9 to 18 in.) in height and can be two-year-old (one-year-one-year) or three-year-old (two-year-one-year) transplants, or alternatively one-undercut-one or one-undercut-one-undercut-one plants. The planting distance may vary from 1.8 to 2.7 m (6 to 9 ft) according to the species.

(ii) The use of larger trees will mean far higher costs both in plants and establishment, and more risk of failure.

(iii) The soil conditions which result from the construction of a main road or motorway often leave much to be desired as far as tree planting is concerned. Embankments may be formed of rock, hardcore and subsoil which are made as firm as possible by rolling in order to produce a strong, stable foundation. Areas adjoining the road may well have had some or all of the top soil removed and heavy road-making equipment may have passed over them many times. These conditions result in intense compaction of what may be, at the best, an infertile soil. Under such conditions conifers are likely to achieve the only worthwhile results, and the establishment of hardwoods may have to be postponed for an appreciable number of years, until soil conditions have improved.

(iv) It is possible to bring good top soil on to the site and to dig this in wherever trees are to be planted, but this will mean a considerable expense which may not be acceptable.

(v) Where the planting site is very compacted the use of mechanical equipment to bore holes may be both economical and produce better growing conditions for the young trees. These excavators are commonly known as 'post hole diggers' and are either mounted on the back of a tractor in the case of the larger patterns, or driven by a chain saw engine, in which case they can be carried and operated by two men.

(vi) Where it is impracticable to plant trees, the prudent use of shrubs can improve the appearance of roadside wastes and at the same time reduce the cost of grass cutting.

CHAPTER X

Trees and Urban Development

1 *The problems of building development*

When land is developed for building purposes several problems arise as regards trees, and some of these are considered in this section.

(a) *The cost of building land*
  (i) At the present time (1989) there is a tremendous demand for land for which planning permission has been granted so that it can be used for building purposes. Since the price of such land is now extremely high, builders and developers are anxious to use as much of it as possible for building, so that tree planting has a low priority.
  (ii) On the other hand, when planning permission is granted, a condition is often laid down as regards the layout and planting of trees.
  (iii) There are consequently two aspects of the matter. On the one hand there is the land developer, who, having paid a high price for the land, consequently wishes to make the most of it, and on the other hand there is the planning authority, who are anxious to obtain an acceptable layout. In many cases this will include the planting of trees.

(b) *Unsuitable trees*
  (i) Not infrequently land which is ripe for development has previously been agricultural land situated on the margin of urban areas. Land of this kind often has large trees growing on it, which may have formed spinneys or hedgerows. As often as not, this will comprise such species as oak, ash and elm.
  (ii) These trees may be unsuitable for inclusion in the new urban environment for a number of reasons such as age, size, species, condition and so on.

2 *The future of existing trees*

(*a*) *Removal or retention*

(i) One of the first decisions to be taken is whether the existing trees shall be retained or whether they should be felled and replaced by others which may neither be of the same species nor ultimately attain the same size.

(ii) The immediate reaction will probably be to retain as many trees as possible, but before deciding to do so, there are several points which should be considered. These will include the following:

(1) Are the trees too old to retain? That is to say, is their expectation of life so short as to make their retention impracticable?

(2) Are the trees to big to retain? Are they of such a size that they might constitute a danger or seriously affect the best form of development from a planning point of view?

(3) Are the trees sound enough and in a sufficiently good state of health to warrant their retention? Trees which are in poor health, are seriously infected by fungus or have been badly damaged, for example by lightning, are very doubtful propositions.

(4) Are the trees of a species which will be acceptable as urban trees when they may in fact be rural trees? The fact that poplars can damage the foundations of buildings renders them undesirable in urban areas, whereas they may be acceptable in open country.

(5) Will any of the trees which are growing on the site interfere with the rights of light of any of the properties?

(6) Are the roots of any trees which are to be retained likely to cause damage to any adjacent buildings?

(7) Will any existing trees which are retained fit in with the proposed development, and will they add to the amenity of the area? Are they in fact the right trees in the right place?

(*b*) *When trees are retained*

(i) Having decided to retain certain trees, steps should be taken immediately to protect them until all building work on houses and roads has been completed.

(ii) In the first place a survey should be made and a plan prepared showing the location of the trees which are to be retained. The trees should also be marked so that they may be identified on the ground. Marking should be done in such a way that it does not injure the tree and can be removed after the site has been developed. The question of site surveys is dealt with in Section 4 of this chapter.

(iii) All those concerned directly or indirectly with the site should be given a copy of the plan and information as to how the trees are marked. Instructions must be issued to ensure that the trees retained are not damaged by lorries, excavation, installations of electricity, gas or telephone services, or building operations generally.

(iv) Where trees are growing in small groups and it is decided to reduce these to a few specimens, care must be taken to ensure that those retained are not subsequently blown down. Generally speaking the trees within a clump or group are less wind-firm than those on the perimeter. If the trees growing on the outside of the group are removed, there is a considerable risk of those which have been growing in the centre, and protected by the outside trees, being blown down. The case of *Sheen* v. *Arden* (1945) was decided on this particular point (see Chapter XV).

(c) Tree Preservation Orders

Tree Preservation Orders are dealt with in detail in Chapter XVII. They are mentioned under this section mainly to emphasize that in certain circumstances it will be necessary to consider whether an Order should be applied for at this stage. Postponement may prove to be unwise.

3 Types of building sites in relation to trees

Land which becomes available for building may have various origins, and some of these are considered below.

(a) Urban areas

This covers what may be described as truly urban land, which has been within the area of a town for some years. This may comprise disused allotments, the gardens of large houses since demolished and areas previously acquired for road widening which have subsequently become redundant. On some of these areas trees may already be growing, more particularly in old gardens, but on others, such as allotments, it is unusual to find any trees of consequence.

(b) Agricultural land

Where building development occurs on the outskirts of towns, it will normally be on land which is, or until recently has been, farm land.

Any trees growing on such land will generally be hedgerow trees or those growing in small spinneys or near farm houses and buildings.

The species will commonly be oak, ash, common elm, wych elm, sycamore or beech, according to that part of the country where the land lies. In some cases such species may be acceptable for towns, but in some cases they will not be.

(c) Woodland

During recent years there has been a movement towards building houses in woodland areas, especially in the Home Counties. Where this occurs it is not so much a case of deciding what trees shall be retained but rather what trees shall be removed. Much of this kind of development has taken place in birch–Scots pine woodlands.

(d) Parkland

As urban development creeps forward across the countryside, country houses and their surrounding park land are sometimes engulfed. When this occurs it is not unusual for large specimen trees, groves, shelter belts and avenues to be swallowed up by the advancing bricks and mortar. In such cases great opportunities are often provided whereby these arboricultural features can be incorporated in the new layout so as to add character to it.

(e) Derelict industrial sites

These areas are generally the most difficult to deal with. The site may comprise slag heaps, tips of industrial waste or 'made' ground, so that conditions for tree planting are not very favourable. The whole question of planting on industrial waste is covered by Chapter XI.

4 Surveys of development sites

As stated earlier in Section 2(b) of this chapter, one of the first steps to be taken with reference to trees on a building site, is to make a survey of the area. The purpose of this is to collect particulars concerning the site itself, and any trees growing on it, and to record any other facts which may have a bearing on the matter. The following are some of the points which should be included in the report. Others may be necessary where additional information is required.

(a) The site

(i) *Location*

Information as to where the site is, with, if necessary, a small-scale plan.

(ii) *Area and extent*
This should include a large-scale plan on which the areas and boundaries are shown. If it is a large area, a 1:10,000 or 1:10,560 Ordnance Survey sheet is generally the most suitable but for small areas, a 1:2,500 Ordnance sheet is best. These can be obtained from the Ordnance Survey through one of its agents.

(iii) *General condition*
A description of the site in broad terms.

(iv) *Soil*
It is important to describe the soil and subsoil and give some indication of depth.

(v) *Elevation*
The altitude or height above sea level of the site should be stated. It may well be necessary to give several heights where noticeable differences in levels occur.

(vi) *Aspect and exposure*
On some sites, these can be limiting factors. The prevailing wind should also be noted under this heading.

(vii) *Rainfall*
The average annual rainfall over the past ten or more years should be included.

(viii) *Drainage*
Information regarding the natural drainage of the site, and details of any streams or watercourses should be recorded.

(b) *Trees on the site*
(i) *General description*
This should describe what categories of trees are growing on the site. These may be divided as follows:
Single or individual trees
Ornamental or common trees
Hedgerow trees
Groups, clumps or groves
Shelter belts
Woodland or spinneys.

(ii) *Species of trees*
This can be conveniently divided into broadleaved trees and conifers.

(iii) *Age*

Where there are a large number of trees on the site, there may be a considerable variation in ages.

(iv) *Size*

This should include top height, girth and crown spread.

(v) *Condition*

In reporting on the condition of an individual tree the following points should be borne in mind:

Health and general appearance

Injuries or damage including presence of fungi, loss of limbs, cavities, die-back of branches or stagheadedness

Estimated remaining life of tree

Whether considered to be a safe or dangerous tree.

(vi) *Position of trees in relation to the proposed development of the site*

This will include:

Value of the tree or trees to the amenity of the site

Proximity to proposed or existing roads and buildings

Possibility of trees causing obstruction when the site is developed.

Need for shelter, wind protection or screening.

(vii) *Legal considerations*

Whether the trees on the site are affected by a Tree Preservation Order or any restrictive covenants.

Whether there are any trees on adjoining land which are so affected.

Whether any woodlands in the vicinity are included in the Woodland Grant Scheme, the Farm Woodland Scheme or the Set-aside Scheme.

(viii) *Recommendations as to retention or removal of the trees on the site*

Any recommendations should be prefaced by a summary of the report and should give reasons for any proposals.

5 *Matters affecting trees on development sites*

The following are some of the matters which can affect trees growing on sites which are to be developed.

(*a*) Care must be taken when carrying out excavation work in the vicinity of trees. If the roots are badly damaged this may result in fungus infection, die-back or windblow.

(b) Work on the site can produce changes in contours and surface levels through excavations, making up ground and road building.

(c) The existing natural drainage of the site may be seriously affected or substantially altered.

(d) When land has been fully developed a large part of it becomes covered with impervious materials which prevent rain being absorbed into the soil. This condition is produced by the erection of buildings, the construction of roads, car parks, playgrounds and so on. Rainwater falling on these areas is removed by drains and sewers. The effect of this is to produce drier soil conditions than existed previously.

(e) Steps must be taken to ensure that adequate space for the growth, of both roots and crown, is provided for existing trees or those which are to be planted.

(f) The erection of large high buildings can have a considerable effect on trees growing in close proximity. Such buildings may tend to produce wind currents and to funnel the wind in certain directions. Trees can also be overshadowed by tall buildings, and if planted too close, their crowns can become unbalanced as they endeavour to grow away from the building.

(g) During building operations the soil on the site often becomes heavily compacted to the detriment of tree growth. If this cannot be avoided, steps must be taken to restore the soil before tree planting takes place.

(h) As work on a building site proceeds it is not unlikely that liquids or materials which are obnoxious to tree growth will be discharged on to the land. This may include waste oil or diesel oil from contractors' plant, creosote, paint and hardcore. Trees must not be planted on ground where such contamination has taken place. If it is essential that they should be, all affected ground must be removed and replaced by fresh clean soil.

(i) When all constructional work has been completed, steps should be taken to clean the whole area and to make good the surface. This will include the removal of plant and machinery, temporary buildings, equipment and builders' rubbish. When this has been done, it may be necessary to cultivate some areas and to carry out a final grading of the surface.

Note. Further information will be found in the following leaflets which are published by the Arboricultural Association:

Trees on Development Sites by D. R. Helliwell (1985).

Tree Survey and Inspection (1988).

CHAPTER XI

Planting Industrial Waste

1 *Classification of industrial waste*

(*a*) In this chapter the term 'industrial waste' is used to cover the workings and unwanted material which result from those undertakings which extract and remove material from the earth. These are set out below and are considered in detail later in this chapter (Section 4). Although they do not come within this definition, refuse tips have also been included, as it is considered that this is the most appropriate chapter in which to place them.

(*b*) Industrial waste may therefore be produced through the following operations:

Colliery tips and slag heaps	China and ball clay workings
Quarry workings and tips	Cement quarries
Sand and gravel pits	Refuse tips
Ironstone workings	

(*c*) In recent years open-cast working of coal and clay have been developed so that in any of the above undertakings there can be two aspects of the work; the restoration of the worked area or 'hole' and the treatment of the excavated waste material.

2 *Objects in planting*

Industrial waste is usually planted for one or more of the following reasons:

(*a*) Amenity
To improve the appearance of the waste area, and as far as possible to convert it from an eyesore into something which is pleasing to look at.

(*b*) Screening
To provide a screen which will hide an otherwise unattractive object. This is also carried out in the interests of amenity.

123

(c) *Surface stabilization*

Where the surface of a tip is likely to be scattered by the action of the wind or washed away by rain, it may be advisable to plant it in order to stabilize the surface.

(d) *Timber production*

In a number of cases old tips and slag heaps have been planted with timber trees and in due course these will require thinning and will produce timber. It is probable that in most cases where industrial waste has been planted timber production has not been the primary object. In such cases, however, as well as improving the amenity of the site, a useful crop of timber has been grown.

3 *General considerations*

(a) *Types of waste*
  (i) The various operations which produce industrial waste are given in Section 1 of this chapter. It will be appreciated that in each case the type of waste produced will vary according to the undertaking. Thus coal mines will produce a different kind of waste to china clay mines or stone quarries.
 (ii) Waste may consist of colliery slag, stone, subsoil, marl, gravel washings, overburden or material unsuitable for the required product, e.g. impure china clay, inferior brick earth and material unfit for the manufacture of cement.
(iii) In some cases such as material from collieries, this may have burnt through spontaneous combustion, and this fundamentally alters its constitution.

(b) *Profile of the site*
  (i) In the case of excavated sites the profile will be that of a pit or hole which will vary very considerably in width, breadth and depth, according to the extent of the bed of material which has been removed.
 (ii) The excavated material is usually tipped so that it forms a heap. This may be irregular in outline or roughly in the form of a cone, an elongated ridge or 'cascade'.

(c) *Water table*

Where the natural water table is high, excavated workings will generally become flooded. This occurs very commonly in the case of gravel pits situated in the vicinity of rivers such as the Thames valley.

4 *Planting of industrial waste*

(a) *Colliery tips and slag heaps*

(i) These may be composed of three distinct kinds of waste. Firstly, rock and shale excavated during the driving of headings to the coal face. This is known variously as spoil, dirt, gob and so on; secondly, material which has become mixed up with coal during mining, and is referred to as slag or slack; and thirdly, slag which has been burnt through spontaneous combustion. This is sometimes referred to as red ash.

(ii) The composition of the types of waste described above may vary considerably, and consequently it is unwise to be dogmatic as regards the species to be recommended for such sites.

(iii) The following will generally be suitable for planting on these sites: Corsican pine, lodgepole pine, Scots pine, Japanese larch, alder, birch and oak.

(b) *Quarry workings and tips*

(i) As in the case of colliery tips, the conditions for tree growth produced by quarry workings and their accompanying waste will depend to a large extent on their origin. Moreover there may be a considerable disparity as regards their texture, which may comprise rejected stone, chippings, washings or overburden.

(ii) The species which will usually grow on quarry workings are: Corsican pine, the larches, alder, beech and oak. Scots pine may also be successful.

(c) *Sand and gravel pits*

(i) Where sand and gravel are worked, the result is generally a pit, and there is normally little or no spoil which will give rise to a heap or tip. Any waste or overburden is usually discharged into a previously worked pit.

(ii) Pits which result from the winning of sand or gravel can be classed as dry or wet according to whether they become filled with water or not. Wet pits normally occur where there is a high water table and this is generally the case in broad alluvial valleys through which a river flows.

(iii) Where the water table is high, the species which will grow readily along the margins of worked pits and in the close surroundings are poplars, willows, alder and lodgepole pine. On the rather drier sites birch, Corsican pine and Scots pine can be established with little difficulty. Where a more ornamental effect is required a large number of trees and shrubs are available such as the Swamp cypress (*Taxodium distichum*), ornamental willows, dogwood, mountain ash and *Buddleia sp.*

(*d*) *Ironstone workings*
(i) In certain parts of the country, more particularly near Corby in Northamptonshire, ironstone has been worked under the open-cast system. This involves the removal of the overlying soil layer or overburden and the excavation of the underlying ironstone. The work is carried out by means of very large mechanical excavators which deposit the overburden in massive ridges. This produces a series of roughly parallel rows with 'valleys' in between which is known as 'hill and dale'.
(ii) The restored soil conditions vary from sand and limestone to heavy clay and the species adopted for planting vary accordingly. On the lighter soils Japanese and hybrid larch, sycamore and Scots pine are the most successful, and in some cases have been planted as a mixture. On the heavier soils alder, oak and poplars have been used successfully.

(*e*) *China clay and ball clay working*
(i) Clay is both mined and worked under the open-cast system. In either case deposits of waste material or overburden are produced, and in the case of open-cast workings large pits are produced which inevitably become water-filled.
(ii) Comparatively little work has been done in planting these areas, but the conditions which are produced have much in common with gravel pits except that the soil structure is substantially different.
(iii) Tree species which can be used for planting these sites include alder, birch, oak, Norway spruce, Sitka spruce and Corsican pine.

(*f*) *Cement quarries*
(i) Since the main constituents of cement are derived from chalk or limestone, the resulting workings are found on these geological formations.
(ii) In many cases, quarrying is very extensive both in depth and extent, and except for the removal of the overburden very little waste is produced. Unless the workings are restored by filling, the excavated areas are not easy to deal with from the point of view of tree planting. Very often planting is limited to the 'floor' of the quarry or pit.
(iii) Species selected for planting must be those which will find chalk and limestone acceptable. Suitable species are Austrian pine, Western red cedar (*Thuya plicata*), beech and sycamore.

(g) *Refuse tips*

(i) The planting of refuse tips is not a practical proposition unless the refuse has been covered by an adequate depth of soil.

(ii) Tips of this kind are often set alight or catch fire through spontaneous combustion, while the formation of methane gas by the refuse can be very dangerous. In such cases, it is essential that any fires are completely extinguished before the site is covered with soil, since burning can continue beneath the surface.

(iii) The choice of species will depend on the soil covering but Corsican, Lodgepole or Scots pine are generally worth considering.

Note. Further information on planting industrial waste will be found in the following Forestry Commission publications:

Bulletin No. 65 – *Advances in Practical Arboriculture* (1987):
'Reclamation of Mineral Workings to Forestry' (pp. 38–41)
'Reclamation: Colliery Spoil' (pp. 42–51).

Research and Development Paper No. 136 – *Tree planting in colliery spoil*.

Arboriculture Research Notes:
Reclamation of Surface Workings for Trees
I – *Landforms and cultivations* (37/84/SSS)
II – *Nitrogen Nutrition* (38/84/SSS).

CHAPTER XII

Hedges, Screens and Shelter Belts

1 *Hedges*

(a) *The use of hedges*

Hedges are planted for various purposes and their uses may be summarized as follows:

(i) *Boundary markers*

To provide a long-lasting means of marking the boundary between two properties or parcels of land.

(ii) *Stock enclosers*

To enable farm stock to be confined within a specified area.

(iii) *Shelter*

To provide shelter for gardens, orchards and stock.

(iv) *Internal partitions*

To act as internal partitions within a garden or nursery either for amenity or to give very local protection against wind or frost.

(v) *Miscellaneous*

Hedges can also be used for various other purposes such as marking the positions of the foundations of walls of ancient buildings, in the creation of mazes and, in their dwarf form, for edgings in gardens.

(b) *Types of hedges*

(i) *General*

Hedges may be classified according to the purpose for which they are grown and broadly speaking there are two main types of hedges: garden and farm. Garden hedges can be further sub-divided regarding species of which they are composed.

(ii) *Garden hedges*

(1) *Evergreen*

The following are some of the species which can be used for evergreen garden hedges:

Atriplex halimus – Tree purslane
Aucuba japonica – Spotted laurel
Buxus sempervirens – Common box
Chamaecyparis lawsoniana – Lawson cypress
Cotoneaster simonsii
Cupressocyparis leylandii – Leyland cypress
Euonymus japonicus – Japanese spindle
Griselinia littoralis
Ilex aquifolium – Holly
Laurus nobilis – Laurel
Ligustrum vulgare – Common privet
Lonicera nitida
Prunus lauroceras – Cherry laurel
Prunus lusitanica – Portugal laurel
Quercus ilex – Holm oak
Rhododendron ponticum
Taxus baccata – Yew
Thuya plicata – Western red cedar.

Those marked with an asterisk are especially suitable for planting near the sea coast.

(2) *Deciduous*

Some of the deciduous species which are suitable for hedges are given below:

Carpinus betulus – Hornbeam
Fagus sylvatica – Beech
Hippophae rhamnoides – Sea buckthorn
Prunus cerasifera – Myrobalan.

(3) *Flowering*

Under this heading are included species which have been selected more for their flowering ability than their hedging qualities. In some cases they tend to spread and require more space than that which is afforded the more usual kind of hedge:

Berberis darwini
x stenophylla
Ceanothus thyrsiflorus
Cytisus scoparius (and cultivars)
Deutzia scabra
Escallonia langleyensis
Forsythia spp.
Pyracantha spp.
Rosa spp.

(iii) *Farm hedges*
  (1) The requisites of a farm hedge are that it should:
    Be stockproof, i.e. provide a strong dense barrier.
    Provide shelter, if necessary.
    Be hardy and quick-growing.
    Withstand cutting.
    Not require undue attention.
    Not be poisonous to animals.
  (2) The following species will form satisfactory farm hedges, but in practice many hedges found on farms consist of a mixture of different species:
    Beech.
      Not very stockproof unless grown on top of a bank.
    Blackthorn.
      A good hedge but slower growing than hawthorn.
    Hawthorn.
      Probably the best all round farm hedge.
    Hornbeam.
      Similar to beech as a stock fence.
    Holly.
      Forms a good hedge but slow growing.

(iv) *Fire blight*
  It should be borne in mind that hawthorn, ornamental thorns, and certain other species are subject to infection by Fire blight (*Erwinia amylovora*), and their use as hedging plants may have to be restricted in some areas. Further information on this disease will be found in Chapter V.

(c) *Planting*
 (i) *Methods*
  Hedges may be planted either on the flat, on top of a bank, or less commonly, at the foot of a bank.
  In the case of farm hedges, a ditch is often dug besides the hedge or bank. It is always wise to ensure that the ground is dug well before planting.

(ii) *Number of plants required*
  (1) When establishing a hedge it is usual to put in a double row with the plants staggered, i.e. those in the first row opposite the spaces between those in the second row, with the exception of holly which should be planted in a single row.

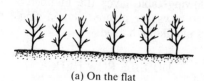

(a) On the flat

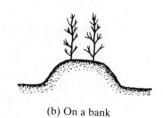

(b) On a bank

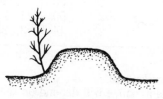

(c) At the foot of a bank

Figure 12.1 Planting hedges

(2) The number of plants required depends on the distance apart at which they are planted and whether there are one or two rows.

(3) The planting distance will depend on:
The species used
The size of the plants
The soil and site conditions
The purpose for which the hedge is planted.

(4) In the past, hedging stock was often planted close together but, with the increase in the cost of plants and labour, the tendency is to extend the distance where conditions are satisfactory. For species such as beech, hawthorn and Lawson cypress, which are 30–46 cm (12–18 in.) tall, the distance in the rows can be 46–61 cm (18–24 in.) and 46 cm between the rows.

The following table gives the number of plants required when two rows are planted. For a single row hedge the numbers should be halved.

DOUBLE ROW HEDGE

Distance apart of plants in the row		Number of plants required per			
cm.	in.	Yard	Chain	Mile	Kilometre
15	6	12	264	21,120	13,115
23	9	8	176	14,076	8,741
30	12	6	132	10,560	6,557
46	18	4	88	7,038	4,370
61	24	2	66	5,280	3,278
76	30	2	50	4,224	2,623
91	36	2	44	3,520	2,185

(d) Tending and maintenance

The tending and maintenance will depend on the species of which the hedge is composed, and also whether it is a garden hedge or farm hedge.

(i) Weeding

After planting, steps should be taken to ensure that the young plants are not overwhelmed by weed growth.

(ii) Draining and ditching

It is essential that a hedge does not become waterlogged and that the young plants are adequately drained. If draining or ditching is necessary, it should be carried out before planting rather than afterwards.

(iii) Pruning and trimming

Garden hedges require clipping or pruning if they are to be kept within bounds and present a neat and tidy appearance.

Farm hedges should be cut and laid in the Leicestershire fashion to produce a stockproof fence. The mechanical hedge cutter will produce a compact, tidy hedge provided it is used regularly on well established hedges.

Note. For more detailed information on hedges, reference should be made to the following. The first, published by the British Trust for Conservation Volunteers, is most explicit and is illustrated by a large number of excellent drawings.

Hedging, by Alan Brooks, revised by Elizabeth Agate (British Trust for Conservation Volunteers, 1984).

Hedges for Farm and Garden, by J. L. Beddall (Faber & Faber, 1950).

Hedges, Shelterbelts and Screens, by A. D. C. Le Sueur (Country Life, 1951).

2 Screens

(a) General

(i) Tree screens may be regarded as being half way between a hedge and a shelter belt.

(ii) Compared with hedges, they are taller, rather wider and frequently are neither topped nor trimmed.

(iii) At some point the differentiation between a hedge and a screen becomes somewhat obscure. The famous beech 'hedge' of Meikleour between Perth and Blairgowrie is a good example of this.

(b) The purpose of screens

Tree screens have two main purposes. Firstly to obscure what may be considered to be the less attractive features of the landscape such as buildings, railway lines, roads, quarries and so on. Secondly to create a sense of privacy by circumscribing a particular area.

(c) Types of screens

(i) Broadly speaking, tree screens can be of two kinds: those formed of one species and those formed of several.

(ii) Screens which are planted in towns are usually not so wide or tall as those found in rural areas.

(iii) It is not always necessary for a screen to be continuous. A satisfactory screening effect can be obtained by planting trees in suitably sited clumps or groups.

(d) Trees suitable for screens

(i) In deciding which trees are most suitable for screening purposes, much will depend on the space available. Where this is unrestricted, large, spreading trees may be planted, but where space is limited smaller and more compact trees must be used.

(ii) *Large trees*
Austrian pine
Beech, especially the fastigiate forms such as Dawych beech
and *var. pyramidalis*
Corsican pine
Hornbeam
Lawson cypress
Leyland cypress
Lime
Norway maple
Poplars, especially Lombardy poplar
Western red cedar

(iii) *Small trees*
Mountain ash
Myrobalan
Whitebeam
Willow

3 *Shelter belts*

(*a*) *Classification*
Shelter belts may be divided into four main classes:
   (i) Belts of trees which afford protection to a farm or area of land
       from the prevailing winds.
  (ii) Groups or clumps of trees which provide shelter to a more
       limited area such as a group of buildings.
 (iii) Localized wind breaks which provide shelter to market
       gardens, orchards, hop gardens and so on.
 (iv) Artificial wind barriers such as walls and interwoven fences.

(*b*) *The advantages of shelter belts*
The successful establishment of natural shelter belts will produce
the following advantages:
   (i) The provision of shelter to livestock during inclement weather.
  (ii) The protection of growing crops.
 (iii) The reduction of soil erosion in those areas which are subject to
       wind, as for example in the Fens.
 (iv) The resultant increase in temperature of the immediate
       surroundings and the consequent effect on stock and crops.
  (v) The provision of nesting sites and cover for game and wild-
       life.

(c) The drawbacks of shelter belts
If shelter belts are to be established, certain drawbacks must be accepted and these may be summarized as follows:
  (i) Shelter belts require a certain amount of land, which will consequently be denied to other uses.
 (ii) Arable crops in the immediate vicinity of shelter belts may be shaded or affected by turbulence resulting from the belt.
(iii) Roots spreading outwards from a belt into the adjoining land may enter into competition with the adjacent crop and reduce the water content of the soil which is in close proximity.

(d) The need for shelter belts
Shelter belts will be found to be of the greatest value where their presence reduces the local wind velocity to such an extent that the general environment is improved. The following are some of the situations where shelter belts are of particular value.
  (i) At the higher elevations such as hill farms, on moorland and on downs.
 (ii) On wide expanses of flat land where wind protection is non-existent, as in many parts of East Anglia and the Fens.
(iii) In coastal areas where protection is required from gales and salt-laden winds.
(iv) On light soils which are subject to erosion by wind.

(e) Establishment of shelter belts
  (i) *Width*
      Shelter belts should not be less than 20 m (66 ft or 1 chain) wide, and a width of 30 m (100 ft or $1\frac{1}{2}$ chains) is often to be preferred. Belts can be wider than this, but the maximum is probably 50 m (165 ft or $2\frac{1}{2}$ chains). Other things being equal, wider belts will produce more thinnings and will usually hold more pheasants.

 (ii) *Planting distance*
      Trees should be planted 2 m (6 ft 6 in.) apart except that the four rows on the exposed side should be at a distance of 2.4 m (8 ft.). This is to ensure ample space for the development of the roots and branches of the outside and therefore the more exposed trees. Planting should be done with great care so as to give the young trees a good start.

(iii) *Grants*
      (1) Grants are available for planting and protecting shelter belts and shelter hedges and for their maintenance during the first three years.

(2) Further information can be obtained from:
   a. The Agricultual Development and Advisory Service (ADAS) of the Ministry of Agriculture, Fisheries and Food, Great Westminster House, Horseferry Road, London SW1P 2AE, or from the Divisional Offices of the Ministry (addresses can be found in the local telephone directory).
   b. The Department of Agriculture and Fisheries for Scotland, Chesser House, 500 Gorgie Road, Edinburgh EH11 3AW.

(iv) *Species*
   (1) *General conditions*
      Broadleaves

Alder	Lime
Beech	Oak
Birch	Poplar
Hornbeam	Sycamore.

      Conifers

Austrian pine	Scots pine
Corsican pine	Japanese larch
Lodgepole pine	Sitka spruce.

   (2) *Sea-coast areas*
      Broadleaves

Abele or White Poplar	Sycamore
Rowan	Whitebeam.

      Conifers

Austrian pine	Monterey pine
Corsican pine	Monterey cypress
Maritime pine	Leyland cypress.

(*f*) *After-care*
(i) At about 15 years the belt will probably require thinning, and this will be repeated at intervals of about five years according to the growth.
(ii) As thinning proceeds and the number of trees is reduced, it is wise to provide an understorey to maintain the efficiency of the belt. This can be done by planting such species as rhododendron and rowan, or underplanting with a shade-enduring species such as Lawson cypress or beech.

Note. For additional information on shelter belts reference should be made to the following:
   Hedges, Shelter Belts and Screens by A. D. C. Le Sueur (Country Life, 1951).

Shelter Belts and Windbreaks by J. M. Caborn (Faber & Faber, 1965).

Trees as a Farm Crop by E. G. Richards, J. R. Aaron, G. F. D'A. Savage and M. R. W. Williams (BSP Professional Books, 1988).

CHAPTER XIII

Hedgerow Trees

1 *General*

(*a*) Hedgerow trees are dependent for their existence on hedges, and where there are no hedges, there will be no hedgerow trees. Even so the presence of a hedge does not necessarily mean that there will be trees growing in it. This may be due to various causes, but the widespread use of the mechanical hedge cutter, which tends to prevent the natural growth of young trees in a hedge, is one of the main reasons.

(*b*) From 1950 to 1985, farmers became more and more concerned with increasing productivity and efficiency. With these objects in view, many farm hedgerows were removed for two main reasons:

(i) To form larger fields which are more suitable for present-day mechanized farming.

(ii) To save the money which would otherwise be spent in cutting and maintaining the hedge.

(*c*) Recently considerable publicity has been given to the removal of hedges by those persons or bodies who are opposed to such action. These complaints tend to overlook two important facts:

(i) Although enclosures of agricultural land had started in Tudor times, the movement reached its climax with the Enclosure Acts of the 18th and 19th centuries. The hedgerow is thus a man-made fence of comparatively recent origin.

(ii) There are still large areas of England in which there are few or no hedges. Such areas include the Fens and parts of Cambridge-shire, Norfolk, Suffolk and Kent.

(*d*) There is a case for the retention of hedges and for their removal, and this is considered in the next two sections.

2 *The advantages of hedgerow trees*

The following is a summary of some of the advantages of hedgerows and hedgerow trees:

(*a*) The provision of shade and shelter for stock.

(*b*) The prevention or reduction of soil erosion by wind.

(*c*) The provision of nesting sites and cover for game and other birds.

(*d*) The establishment of a supply of timber either for use on the farm or as a financial reserve.

(*e*) The very considerable contribution towards the amenity and beauty of the country.

3 *The disadvantages of hedgerow trees*

The disadvantages of hedgerow trees may be summarized as follows:

(*a*) The interference caused by tree roots when permanent pasture is ploughed up. This may continue for two or three years until the roots have been worked out.

(*b*) The removal of moisture and plant food from the surrounding land by the roots of hedgerow trees.

(*c*) The causing of damage to field drains and walls where they exist in close proximity to the hedge.

(*d*) The overshadowing of crops and the effect of turbulence on them.

(*e*) The provision of cover for vermin. This must be weighed against the advantages of hedges as cover for game and beneficial birds.

(*f*) The restriction of the use of mechanical hedge cutters.

(*g*) The risk of damage being caused by trees blowing down or by a limb falling. This risk will be highest in roadside hedges but it is possible to insure against this at a reasonable cost.

4 *Species*

(*a*) The *Census of Hedgerow and Park Timber and Woods under Five Acres*, which was undertaken by the Forestry Commission and published in 1951, showed that oak, elm, ash, beech and sycamore were the principal hedgerow trees found in England in the order given. In Wales the order of prevalence was oak, sycamore, elm, ash and beech.

(*b*) Conifers seldom occur as hedgerow trees, although Scots pine, European larch, Norway spruce and the common silver fir are sometimes found.

(*c*) The common elm, on account of its capacity to spread by means of suckers and its ability to produce a large volume of timber in a comparatively short time, is probably the best hedgerow tree. Unfortunately, owing to elm disease, this tree can no longer be considered a feasible proposition.

5 *Methods of establishment*

(*a*) Hedgerow trees are seldom planted artificially but normally grow from seed which has found its way into the hedge bottom through the agency of birds, animals or wind. The exception to this is the common elm, which spreads by root suckers.

(*b*) If it is desired to plant hedgerow trees instead of relying on natural regeneration, or if it is thought necessary to supplement natural growth, there are two courses open:

(i) To plant in the hedge, or
(ii) To plant beside the hedge.

but both of these methods have disadvantages.

(*c*) Planting in the hedge is difficult, impractical and very expensive. Before any trees can be planted, individual positions must be prepared, and this means cutting away part of the existing hedge and then digging a planting hole in the hedge bottom. In well-established hedges this is a formidable task, which on some sites may prove insuperable. If these difficulties are overcome it is then necessary to plant a reasonably large standard tree so that the crown is well above the top of the hedge, and this adds considerably to the cost.

(*d*) The alternative to planting in the hedge is to plant beside it, in the adjoining field or on the adjacent verge. In this way the work, and therefore the cost, is greatly reduced, but the need for protection is greatly increased, as the trees will be clear of any shielding growth. Defence against cattle and farm implements must be provided, and this usually takes the form of a protective fence or possibly tree guards. Where a hedge has a ditch on one side, the trees must be planted on the side furthest from the ditch. Failure to do this will cause difficulties if the ditch is to be cleaned out with a mechanical digger. So as to avoid similar difficulties arising when hedge-cutting machinery is used, it will be necessary to operate the machine from the side of the hedge on which the trees have not been planted.

6 *The alternative to hedgerow trees*

(*a*) Where the cost of planting or maintaining hedgerow trees is considered to be too great, timber can still be grown elsewhere on the farm.

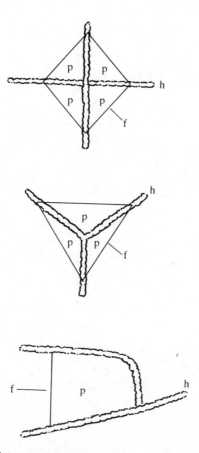

Figure 13.1 Planting around hedge intersections (h = hedge; f = new fence; p = area for planting)

(*b*) The sides of valleys which are too steep to cultivate will generally grow good crops of trees.

(*c*) On the higher and more exposed sites shelter belts may be established, and these are described in Chapter XII.

(*d*) Small spinneys may be established at hedge intersections by fencing between the 'arms' of the hedges (see Figure 13.1).

Note. Further information on hedgerow trees will be found in:

Farm Woodland Management by J. Blyth, J. Evans, W. E. S. Mutch and C. Sidwell (Farming Press Ltd, 1987).

CHAPTER XIV

Avenues, Parks and Amenity Woods

1 *Avenues*

Avenues can provide one of the most attractive forms of tree planting and the following notes provide a guide to their layout and design.

(*a*) *Length*
 (i) If an avenue leads from one definite point to another, e.g. from a house to a monument, its length is automatically defined.
(ii) In other cases, the length should be decided in the light of certain guiding factors, including the following:
   (1) If too long, an avenue tends to become monotonous.
   (2) If very long, care must be taken to ensure that it is not too narrow, in which case it will tend to shrink to vanishing point before the end of it is seen.
   (3) If a very long avenue is made too wide, in order to avoid the last pitfall, the trees of which it is composed will appear to be dwarfed.
   (4) Avenues should not normally be more than one-half to three-quarters of a mile in length if the above points are to be avoided.

(*b*) *Width*
 (i) The width between the two flanking rows of trees may vary considerably, according to:
   (1) The habit of the species planted: large-crowned trees need more space than those with narrow crowns.
   (2) The length of the avenue: long avenues should be rather wider than those of shorter length.
   (3) The number of rows of trees of which the avenue is composed: an avenue of two single rows should be rather narrower than one of two double rows.
(ii) Consequently, it is difficult to lay down the exact width of an avenue, and the following should be regarded only as a guide:

143

Length of avenue		Width of avenue	
metres	yards	metres	feet
91	100	9	30
91–182	100–200	12	40
182–365	200–400	15	50
365–548	400–600	18	60
548–640	600–700	21	70
640–731	700–800	24	80

(c) *Distance between trees in the rows*
  (i) Avenues may be of two types:
      (1) Close-planted: adjacent trees merge into each other.
      (2) Wide-planted: adjacent trees are far enough apart to develop as individuals.

  (ii) *Close planting*
      (1) If unthinned, a close-planted avenue eventually becomes almost a hedge on a large scale.
      (2) It may be adopted for trees which have an irregular outline.
      (3) The distance between trees is from 6.0 m to 9.1m (20 ft to 30 ft).
      (4) By removing alternate trees in due course, a close-planted avenue can assume the appearance of a wide-planted one.

  (iii) *Wide planting*
      (1) By this method each tree in the avenue is allowed to develop as an individual.
      (2) It is best suited to trees which are symmetrical in appearance, and consequently is often used for conifers.
      (3) The distance between the trees is from 9.1 m to 18.2 m (30 ft to 60 ft).

(d) *Species*
  (i) A species suitable for planting in an avenue should:
      (1) Attain a reasonable height.
      (2) Have a shapely outline.
      (3) Develop symmetrically.
  (ii) Species which form ragged or irregular crowns should be avoided.
  (iii) The following species are among those suitable for planting in avenues:

(1) *Broadleaved*:

Beech	Oak
Chestnut, horse	Plane
Lime	Poplar
Norway maple	Sycamore

(2) *Conifers*:

Cedars: Atlas, Deodar, Lebanon

Douglas fir

Larch

Lawson cypress

Pines, especially Scots, Corsican, and the Monterey pine *(P. radiata)*

Redwood (*S. sempervirens*)

Silver firs

Wellingtonia

Western red cedar (*Thuya plicata*).

2 Parks

(a) General

(i) Since trees planted in parks are often allowed to develop as individuals or widely-spaced groups they frequently tend to develop girth and crown at the expense of height.

(ii) Consequently, branch development in many cases is much greater, and in maturity and decline the presence of large branches often results in serious damage by wind, snow, frozen rain, and so on.

(iii) In the management of park trees, the greatest attention should be paid to pruning, the treatment of wounds, and attacks by insects and fungi.

(iv) Several methods can be adopted for planting park trees, of which the following are the more usual:

(1) Single trees   (4) Groves

(2) Groups   (5) Belts.

(3) Clumps

(b) Single trees

(i) These should not be too numerous.

(ii) It is essential that single trees should be shapely and symmetrical.

(iii) Suitable species for this purpose include:

Beech	Cedar of Lebanon
Horse chestnut	Deodar

Lime	Scots pine
Pedunculate oak	Plane.
Atlas cedar	

(iv) Particular care is needed in the pruning of single trees, or they will become badly balanced and misshapen.

(c) *Groups*

(i) Groups may be defined as clusters of three to five trees, the crowns of which unite in a compact mass of foliage.

(ii) The result is to produce a more solid effect than single trees.

(iii) Although more than one species can be planted in a group, it can be more effective to plant only one.

(d) *Clumps*

(i) The objects of a clump are:
   (1) To provide a background where it is needed.
   (2) To obscure or break up a bare skyline.
   (3) To give depth and variety to the scenery.

(ii) Clumps may consist of one or more species, but a mixture will often prevent too formal an appearance.

(iii) Clumps should, if possible, be irregular in shape, so as to avoid too rigid an outline.

(e) *Groves*

(i) These are used where it is desired to divide up an area into several smaller parts, and can be adopted for separating the more formal part of a park from the remainder.

(ii) Groves may occupy such situations as:
   (1) Long ridges.
   (2) High or broken ground.
   (3) Level ground, so as to form short rides or avenues.

(iii) The shape and size vary, but irregular areas are probably best, especially if the margins are broken.

(iv) Any of the species commonly used for woodland planting may be used for groves, subject to site restrictions.

(f) *Belts*

(i) Belts may be planted for shelter, screening, or in connection with shooting.

(ii) For details reference should be made to Chapter XII.

3 *Amenity woods*

(*a*) *Definition*
(i) Amenity woods are woods whose management is primarily based on considerations other than those of timber production, although in fact they may produce timber.
(ii) Such woods include the following:
    (1) Woods planted near the mansion house for beauty, privacy, or shelter.
    (2) Shelter belts.
    (3) Parkland planting for landscape effect.

(*b*) *The management of amenity woods*
(i) The main object in the management of these woods is to maintain a permanent cover of trees.
(ii) Clear felling must be avoided at all costs.
(iii) To ensure permanency there must be an adequate series of age classes; that is to say, there must be trees of all ages or groups of ages on the area so that shelter is continuously maintained.
(iv) Ornamental trees, flowering shrubs, and exotics may be introduced along the margins of amenity woods if it is thought desirable.

CHAPTER XV

Trees and the Law

1 *The ownership of trees*

(*a*) *General*
   (i) Unless there is evidence to the contrary, trees belong to the owner of the land upon which they are growing.
   (ii) However, the ownership of the trees may be separated from that of the land if, for example, the land has been sold and the trees have been retained by the vendor.
   (iii) Furthermore, an owner may be subject to certain restrictions as regards the trees growing on his land, which may, for example, be imposed by:
      (1) A Tree Preservation Order.
      (2) Current legislation, e.g.
         The Forestry Act 1967
         The Trees Act 1970.

(*b*) *Trees on boundaries*
   (i) In deciding the ownership of trees on and near the boundaries of a property, it is necessary to decide in the first place, what and where the actual boundary is. For example, where there is a hedge and ditch, the boundary will be the edge of the ditch furthest from the hedge, unless there is acceptable evidence to the contrary. Thus in Figure 15.1 the boundary between the two properties A and B is indicated by the arrow line, the hedge and ditch belonging to B.
   (ii) If there is no ditch the ownership of the hedge may be:
      (1) Marked on the deeds of the property.
      (2) A party hedge, belonging equally to the adjoining owners.
   (iii) If the owner of a piece of land plants a tree close to the boundary of the adjoining property, the tree belongs to him, even if some of the roots encroach on to his neighbour's land or if some of the branches overhang it (*Holder* v. *Coates* (1927), Moo. & M. 121).

148

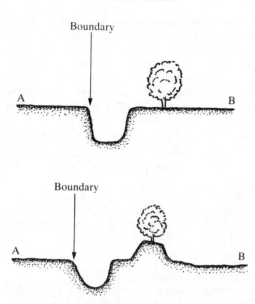

Figure 15.1 Hedge boundaries (see (b)(i), facing page)

(iv) If there is no evidence as to who planted the tree, the question must be judged on the facts of the case. Generally speaking the tree will belong to the owner of the land, on which the major portion of it is growing.

(v) If a tree is growing exactly on the boundary between two properties, it is normally considered to be the joint property of the two owners concerned, unless evidence to the contrary is forthcoming.

(c) *As between landlord and tenant*

(i) The position regarding trees as between a landlord and his tenant will depend on the terms of the contract of tenancy and also in certain cases on whether the trees are legally regarded as timber.

(ii) It is usual for a lease or agreement between a landlord and tenant to contain a clause relating to the trees growing on the property which is let.

(iii) If there is no such clause or no express covenant relating to the trees, the tenant will have a number of rights regarding them and these will include the following:

(1) Right to enjoy shade, shelter and fruit of the trees.
(2) Right to cut underwood, coppice and trees which will throw stool shoots, provided that they are not timber or fruit trees or are trees planted for ornament or shelter or are trees which provide support.
(3) Right to take timber for the repair of his house, if he is responsible for repairs, and for fences, gates and the repair of agricultural implements.
(4) Right to fell and use dead trees so long as they are not sound timber trees or ornamental trees. Windfalls of sound timber belong to the landlord but those of unsound timber belong to the tenant.
(5) In view of the above rights, it is usual for a landlord to reserve to himself all trees growing on the property which is leased. The wording of such reservations varies and may be in the form of 'all timber and timber-like trees' or simply 'all trees'.

(d) *The legal definition of timber*
(i) In the case of *Honywood* v. *Honywood* (1874), 18 Eq. 306, it was stated that the question of what is timber depends firstly on the general law of England and secondly on the special custom of the district concerned.
(ii) By general law, oak, ash and elm of twenty years of age and above are timber. By special custom beech is timber in the counties of Bedford, Buckingham, Gloucester, Hampshire and Surrey. Birch is timber in Cumberland and Yorkshire, while in some districts other species including lime and cherry are held to be timber.
(iii) In some parts, 'timber' is considered to refer to trees of a certain size. From the case of *Whitty* v. *Lord Dillon* (1860), 2 F. & F. 67, it would appear that the minimum size is 6 in. (15 cm) in diameter.

2 *Overhanging trees*

(a) Trees may encroach on to another's property in two ways: by their branches or by their roots. This section is concerned only with trees which overhang an adjoining property or highway. The effect of tree roots entering into adjacent land and causing damage is dealt with in the following section.

(b) If the branches of a tree grow over the boundary between two properties, the owner of the adjoining land may cut them back up to

the line of the boundary (*Earl of Lonsdale* v. *Nelson* (1823), 2 B. &
C. 311).

(*c*) The case of *Lemon* v. *Webb* (1894), 3 Ch. 1, which is one of
the most important cases on the subject of overhanging trees,
established the following points:

(i) An adjoining owner may cut back the branches of an over-
    hanging tree as far as his boundary without giving notice to the
    owner of the tree. However, it would be a neighbourly act if,
    prior to cutting the branches, the owner of the tree was
    informed.

(ii) On the other hand an adjoining owner may not enter the
     property on which the tree is growing in order to cut back the
     encroaching branches, without first giving notice to the owner
     of the tree.

(*d*) The case of *Smith* v. *Giddy* (1904), 2 K.B. 448, established
the right of an adjoining owner to claim damages from the owner of
an encroaching tree, where actual damage occurs. This decision
followed the principle laid down in the basic case of *Rylands* v.
Fletcher (1868), L.R. 3, H.L. 330, that where a person has on his
land something which may harm his neighbour, he must keep it
within his own boundaries. If he does not, and consequently his
neighbour suffers harm, he will be liable for such injury.

(*e*) Where branches or limbs are removed by an adjoining owner,
they remain the property of the owner of the tree. In the same way
any fruit growing on the branches of a tree which overhangs another
property belong to the owner of the tree (*Mills* v. *Brooker* (1919), I
K.B. 555).

(*f*) Where a branch overhangs a highway it may cause damage to
passing vehicles, and the owner of the tree may be sued. The case of
British Road Services Ltd v. *Slater and Another* (1964), E.G. 441, is
concerned with this matter.

3 *Damage by tree roots*

(*a*) The roots of trees can encroach on adjoining property by
growing through the soil and over the boundary. In doing so, there
are two main differences between encroachment by branches and
encroachment by roots:

(i) Branches can be observed with little or no difficulty, roots can-
    not. Branches may obstruct light, shed seeds on to the ground
    beneath them or produce leaves which will block rainwater

gutters. In all such cases any damage and its effect is reasonably apparent to any competent observer.

(ii) Roots may damage foundations or seriously affect drains and sewers. If this occurs it may be impossible to see and difficult to detect except in advanced stages of damage.

(*b*) Generally there are three kinds of damage which are caused by tree roots.

(i) Physical pressure of the roots against walls, footings, paths and so on.

(ii) Blocking of drains when roots succeed in entering them.

(iii) Removal of moisture from the soil which causes shrinkage and consequently subsidence. Shrinkage particularly occurs on clay soils and is aggravated in a dry summer, while expansion occurs during winter when conditions are generally wet.

(*c*) The trees whose roots cause the most damage are poplars, but elm and ash can also give serious trouble. Large shrubs as well as trees can cause difficulties if planted against the walls of a house or building.

(*d*) Tree roots can extend for a considerable distance from the tree itself, and generally speaking the larger the tree, the further are its roots likely to spread.

(*e*) A person whose property is damaged by roots can take the following action:

(i) He can recover damages in respect of the injuries which have been caused.

(ii) He can sever the roots which have encroached on his property, at any point on his own premises.

(iii) He can apply for an injunction to restrain his neighbour from permitting the roots of his neighbour's trees from encroaching upon his property.

(*f*) It should be remembered that if the owner of a property does not cut back the invading roots and such roots cause damage to his property, he still has the right to claim for such damage from the owner of the tree.

(*g*) To cut down an offending tree may not be enough and it may be necessary to kill the roots by an application of chemicals to the stump. This is more likely to be necessary in the case of those species whose roots tend to produce suckers, such as poplars and elms.

(*h*) Since 1940 at least 27 cases have been heard on the matter of damage by tree roots, and these are given in Section 6 of this chapter. In 13 cases, the species concerned were poplars. It is an unfortunate fact that trees which grow to a large size are not suitable for planting in the immediate vicinity of buildings, and these are generally the conditions which occur in towns. This does not refer to planting such trees in parks and other open spaces where they have ample space, but to sites on which houses have been built in close proximity to each other.

(*i*) Where trees are already growing on a property and an adjoining owner decides to build a house near them, it is no defence for the owner of the trees to claim that the adjoining owner built his house at his own risk because he knew that the trees were already there (*McCombe* v. *Read* (1955), 2 Q.B. 429).

(*j*) No distinction can be drawn between trees which have been planted and those which are self-sown as regards the damage they cause. It is no defence to say that damage has been caused by natural growth (*Davey* v. *Harrow Urban District Council*, 2 A.E.R. 305).

(*k*) *Quia timet* action: if an owner fears that something which exists or may be established on the adjoining land will create a nuisance, he can bring an action before the nuisance has begun because of what he fears may happen. This is known as a *quia timet* action and could be for an injunction, damages or both. For example the owner of land adjacent to a row of trees which are growing on his neighbour's property may wish to build a house on his own land, but in close proximity to his neighbour's trees. He may fear that the roots of those trees will damage his proposed house and he may therefore sue the owner of the trees *quia timet*. Whether he would succeed in such action is another matter. The case of *Lemos* v. *Kennedy Leigh Developments Ltd* (1960), 105 SJ 178, is concerned with such an action.

4 *Poisonous trees*

(*a*) Among the commoner trees and shrubs which are poisonous are the following:

Box	Privet
Cherry Laurel	Rhododendron
Daphne (Mezereon)	Spindle
Horse Chestnut	Yew.
Laburnum	

The foliage of yew trees seems to be rather more poisonous when it is dead and the leaves have withered, as occurs in the case of hedge clippings. Virtually all of the law cases dealing with poisonous trees are in respect of yews.

Note. For a detailed account of poisonous plants which are found in the United Kingdom, reference should be made to *Poisonous Plants and Fungi* by M. R. Cooper and A. W. Johnson. This was published by HM Stationery Office for the Ministry of Agriculture, Fisheries and Food in 1988.

(*b*) The law relating to poisonous trees is based on the principle which was established in the case of *Rylands* v. *Fletcher* (1868), L.R. 3, H.L. 330, which is referred to in Section 2. If the owner of a poisonous tree allows it to grow over his boundary so that his neighbour's cattle can reach it, and if they do so and consequently die of poisoning, the owner of the tree is liable. This decision was reached in the case of *Crowhurst* v. *Amersham Burial Board* (1878), 4 Ex. Div. 5.

(*c*) On the other hand it was held in *Ponting* v. *Noakes* (1894), 2 Q.B. 281, that if a horse reaches over the boundary and eats the foliage of a poisonous tree which is growing on his neighbour's land and which does not extend over the boundary, the owner of the tree is not liable.

(*d*) The position as between landlord and tenant is as follows. If land is let to a tenant on which there are poisonous trees, or if access can be gained to such trees from the land let, and if the trees and access were apparent when the tenant took the land, then the landlord will not be liable if the tenant's cattle are poisoned (*Erskine* v. *Adeane* (1873), 8 Ch. 756). Similarly, if a poisonous tree growing on other land belonging to the landlord was overhanging the land let to the tenant *when* he took the land, and he suffers loss, the landlord will not be liable (*Cheater* v. *Cater* (1918), 1 K.B. 247).

5 *Dangerous trees*

(*a*) *Why trees may be dangerous*
A tree may be potentially dangerous for several reasons. It may be a poisonous tree, it may be partly or wholly decayed or rotten and therefore liable to fall, it may be a species which is prone to drop its branches such as elm, or it may be of such a shape or size that it is unsafe under the circumstances in which it is growing. Poisonous trees have already been considered, and in this section the term 'dangerous trees' is used in respect of those trees which, for one reason or another, may fall either as a whole or in part, and in so doing constitute a danger.

(b) Inspecting trees

Before deciding whether a tree is dangerous or not it is necessary to make a careful inspection of it. In doing so, it is advisable to cover systematically the different points which should be looked for when examining a tree. A suggested procedure is as follows:

(i) First of all look at the tree as a whole from a little distance away, walk around it and assess:

General health

Balance of the crown

Extent to which it leans in any particular direction

Physical damage to any branches

Presence of dead branches.

(ii) Next make a closer inspection using binoculars where necessary and note:

Condition of the main stem or trunk

Presence of any holes or areas of rot at the *base* of the tree

Position of any holes in the bole or larger branches

Evidence of any fungus infection, i.e. fructifications (toadstools, brackets etc.), bark falling away or die-back of branchlets

Whether any limbs have broken off leaving areas through which fungi may gain access

Presence of any dead branches

Presence of water pockets in the forks of branches.

(iii) Finally make a detailed inspection of any part of the tree which cannot be seen from ground level (with or without the aid of binoculars). This may require the use of a ladder, ropes and so on, and the safety precautions of such an inspection should not be overlooked.

(iv) The results of such an inspection should be carefully recorded in a notebook. Sketches will often be found most helpful.

(c) Some characteristics of a potentially dangerous tree

A potentially dangerous tree may have one or more of the following characteristics:

(i) An unbalanced crown.

(ii) A trunk which is noticeably growing out of the vertical.

(iii) Dead or dying branches or limbs.

(iv) Fructifications of fungi growing on some part of it.

(v) Holes in trunk or limbs.

(vi) Conspicuous patches of dead bark.

(vii) Stains on the bark due to water continually running down it.

(viii) Be so sited that its roots are likely to have been damaged or

weakened by excavations, e.g. in connection with building operations or road construction.
(ix) Be an elm.

Note. Further information will be found in the following:
Forestry Commission Arboricultural Leaflet No. 1 – *External Signs of Decay in Trees* (1984).
Forestry Commission Pamphlet – *The Recognition of Hazardous Trees* (1988).
Arboricultural Association Leaflet – *Tree Survey and Inspection* (1988).

(*d*) *Dangerous trees and the law*
(i) If a tree or part of a tree falls on to adjoining land or on to the highway, and consequently causes damage or injury to property or to an individual, the general rule is that the owner of the tree will be liable only if it can be established that he has been negligent. (*Caminer and Another* v. *Northern and London Investment Trust Ltd* (1949), 2 A.E.R. 486).
(ii) If it can be shown that the owner of the tree knew or ought to have known that the tree was dangerous and that he took no steps to deal with it, he will be guilty of negligence and therefore liable for any damage or injury which may result (*Kent* v. *Marquis of Bristol* (1940), *Quarterly Journal of Forestry*, January 1947).
(iii) Negligence may be described as omitting to do what a prudent or reasonable man would do, or doing what a prudent or reasonable man would not do.
(iv) From a number of cases concerned with dangerous trees it is clear that although a tree is not in itself a dangerous thing, a landowner has a duty towards others to see that they are not endangered by his negligence. To ensure this, a regular inspection of roadside trees, at least, should be carried out.
(v) As regards the removal of a tree which has fallen on to a highway, it was held in the case of *Williams* v. *Devon County Council* (1966), *Estates Gazette*, 10 December 1966, that if an owner shows that he took reasonable care to see that the tree did not cause an obstruction, the cost of removing it cannot be recovered from him. In this case it was shown that the tree was sound and had been continuously inspected, and this action constituted reasonable care. The case arose over an interpretation of the wording of Section 9(1)(c) of the Highways (Miscellaneous Provisions) Act 1961.

(vi) The question of insurance relating to dangerous trees is dealt with in Section 7 of this chapter.

6 *Law cases concerning trees*

There have been many law cases concerning trees, and brief accounts of some of these are given below. These accounts are believed to represent the true facts of each case, but neither the author nor the publishers can accept any legal responsibility for any action which may arise out of their use. They are placed under the following headings and arranged in alphabetical order under those headings.

(*a*) Overhanging trees.	(*e*) Tree preservation orders.
(*b*) Damage by tree roots.	(*f*) Covenants relating to trees.
(*c*) Poisonous trees.	(*g*) Damage to plantations.
(*d*) Dangerous trees.	(*h*) Definitions.

These are only summaries of the various cases, and reference should be made to the original reports for the full details.

(*a*) Overhanging trees

(i) *British Road Services Ltd* v. *Slater and Another* (1964), *Estates Gazette*, 30 May 1964.

A branch of an old oak, which grew beside a public highway, projected about 2 feet over the roadway, at a height of approximately 16 feet. The plaintiff's lorry was carrying a load of packing cases which reached to a height of 16 ft 4 in. above ground level. When pulling into the side of the road to allow another lorry to pass, a packing case on the top of the load hit the projecting branch, fell into the road and caused another lorry to take such avoiding action that it left the road and was damaged.

It was held that the branch constituted a nuisance because it interfered with the use of the highway. However, the defendants were not liable for such nuisance because firstly, the occupier of land is not liable to remedy a nuisance if he is not aware of it or, with reasonable care, should have become aware of it. Secondly, it could be presumed that the defendants had no such knowledge of the nuisance which could render them liable for failing to remedy it.

(ii) *Hale* v. *Hants & Dorset Motor Services and Another* (1947), 2 A.E.R. 628

The plaintiff was riding on the top floor of a bus belonging to the defendants when it was struck by the branches of a tree which

broke some of the windows, and as a result a piece of glass
entered the plaintiff's eye. The tree in question had been
planted on the side of the highway by the Poole Corporation
under powers provided by the Road Improvement Act 1925,
Section 1(1). The plaintiff brought an action against the bus
company for damages suffered by him through the company's
negligence.

The company denied negligence and said that the accident
was due to the bus hitting the branches of the tree, which
constituted a nuisance and which had been created by the
Corporation who had planted them.

It was held that the bus company and the Corporation were
liable. The defendants appealed but the appeals were dismissed.

It may be noted that Section 1(2) of the Road Improvement
Act 1925 states:

No such tree . . . shall be placed, laid out or allowed to remain in such
a situation as to hinder the reasonable use of the highway by any
person entitled to the use thereof, or so as to be a nuisance or
injurious to the owner or occupier of any land or premises adjacent to
the highway.

(iii) *Lemmon* v. *Webb* (1894), A.C. 1
The plaintiff, who was the owner of some residential property,
sold the adjoining property to the defendant. There were
growing on the boundary some large oak and elm which
overhung the defendant's property and which the defendant
removed without giving notice to the plaintiff.

It was held that it constituted a nuisance to allow the
branches to overhang the adjoining property. At the same time
the owner of such property could abate such nuisance only after
giving notice to the owner of the trees that he intended to do
so. Judgment was consequently given in favour of the plaintiff.

On appeal however it was held that it was *not* necessary to
give notice before cutting back the overhanging branches,
provided that the affected person carried out such cutting on
his own property and did not enter on to the land of the adjoining owner. The appeal was therefore allowed.

(iv) *Lonsdale* v. *Nelson* (1823), 2 B. & C.
In this case it was held that the boughs and branches of trees
which overhang any man's land so as to be a nuisance, may be
cut by the owner or occupier of such land.

(v) *Mills* v. *Brooker* (1919), 1 K.B. 555

The plaintiff owned ten apple trees which grew some 8 feet from the boundary between his land and the defendant's. The branches of these trees overhung the defendant's land, and when they bore a crop of apples the defendant picked the fruit which grew on the branches which overhung his property and sold them. He claimed that he had a right to do this since he had a right to cut back the overhanging branches.

The plaintiff subsequently sued for damages but it was held that although the defendant had the right to remove the nuisance, he could not take the materials which caused the nuisance and appropriate them to his own use. Consequently judgment was given for the plaintiff for 10 lb. of apples.

The defendant appealed but it was held that the apples were the plaintiff's property both before and after severance, and he had the right to possession of them. The appeal was accordingly dismissed.

(vi) *Smith* v. *Giddy* (1904), 2 K.B. 448

The plaintiff occupied a garden, and certain ash and elm trees which grew on the boundary and belonged to the defendant overhung the plaintiff's land and caused some £50 worth of damage to his fruit trees. He therefore brought an action in the County Court, but the Court held that his action did not lie, since his only remedy was to cut back the overhanging branches himself. The plaintiff then appealed and his appeal was granted.

This case is of considerable interest because, until it was decided, it was supposed that a plaintiff's only remedy was to abate the nuisance himself. It therefore established the principle that a plaintiff does not lose his right to bring an action for damages simply because, instead, he could have cut back the branches and thus abated the nuisance.

Summary of species
In the six cases summarized above, the following species were involved:

Apple	1
Ash and elm	1
Oak	1
Oak and elm	1
Unspecified	2

(b) **Damage by tree roots**

    (i) *Acrecrest Ltd* v. *W. S. Hattrell & Partners and Another* (1979), *Estates Gazette*, 15 December 1979

Cracks appeared in the structure of flats belonging to the plaintiffs who sued the architects and Harrow Borough Council for damages. Before the flats were built, a number of elms, poplars and fruit trees were growing on the site which was on London clay. Some of these were removed, and it was submitted that this had resulted in an accumulation of water which would otherwise have been absorbed by these trees. Furthermore that this water had caused the subsoil to swell or 'heave' which produced cracks in the building. This is an unusual case since the damage was held to have been caused by the *removal* of trees growing on the site.

    (ii) *Attfield* v. *Wilson* (1949), *Estates Gazette*, June 1949

The defendant entered into occupation of his house and garden in October 1946. In November the plaintiff pointed out to him cracks which had appeared in his concrete path and yard, and suggested that these were caused by the roots of three poplars which were growing in the defendant's garden near the boundary between the two properties. The defendant agreed to have the trees felled but no action was taken. In May 1947 the plaintiff showed the defendant a poplar shoot near his garage, but no action was taken. During the summer considerable damage occurred to the premises, and since the plaintiff was unable to see the defendant, he wrote to the defendant's father, who owned the property. The trees were then felled.

The defendant maintained that much of the damage occurred before he occupied the premises and further that the plaintiff had failed to take reasonable steps to mitigate his loss. However, it was held that the plaintiff could not have done much more than he did, and judgment was given for the plaintiff.

    (iii) *Bridges and Others* v. *Harrow London Borough* (1981), *Estates Gazette*, 17 October 1981

This case concerned an action by the plaintiff in respect of damage to their property, alleged to be caused by the roots of two oak trees, about 140 years old. It was held that these two trees, which were growing on the roadside, were the property of the defendants, and damages were awarded to the plaintiffs, with costs.

(iv) *Brown* v. *Bateman* (1955), *Estates Gazette*, 5 March 1955

From September 1949 until December 1951 the defendant owned a house next to the plaintiff's. Three elms were growing in the defendant's garden, 12, 16 and 20 feet respectively from the plaintiff's house. In 1947 cracks appeared in the wall adjacent to these trees, and as the cracks became worse complaints were made to the defendant, who took no action in the matter. The trees were ultimately felled in 1952 by the new occupier after the defendant had sold the property.

Damages of £292 were claimed by the plaintiff but he was awarded only £50 since it was held that the defendant was liable only from the date on which the plaintiff informed her of the damage until she disposed of the property.

(v) *Bunclark and Others* v. *Hertfordshire County Council and Another* (1977), *Estates Gazette*, 30 July 1977

The plaintiffs brought this case on the grounds that their flats had been extensively damaged by the roots of various species of trees growing on the defendant's land. The site was on London clay and it was submitted that the roots, by abstracting moisture from the soil, had caused it to shrink. This had resulted in damage to the buildings and judgment was given for the plaintiffs with costs.

(vi) *Butler* v. *Standard Telephones and Cables Ltd* (1940), 1 K.B. 399

In 1930 the defendants planted a number of Lombardy poplars on their property between 10 and 13 feet from the plaintiff's houses. Four years later a considerable amount of settlement occurred but the cause was not known. In an attempt to overcome the trouble the walls of the houses were underpinned.

In 1937 further settlement took place and the collapse of the buildings seemed likely. Judgment was given for the plaintiff, who claimed damages for trespass and nuisance, and he was awarded £450 damages.

(vii) *Catell* v. *Bedingfield* (1946), *Estates Gazette*, 9 February 1946

Twelve poplars grew on the defendant's land and the plaintiffs alleged that damage was being caused to water pipes, a concrete path, an underground electric supply line and the roof and chimneys of the plaintiff's house by the roots and branches of these trees.

The plaintiffs admitted that two of the trees had been felled by a workman on their instructions and the defendant's counter-claimed damages to the amount of £50 and at the same time denied liability. This counter-claim was dismissed, and damages in the sum of £18 were awarded to the plaintiffs. An injunction was also granted restraining the defendants from allowing their trees to damage the plaintiff's property.

(viii) *Coupar* v. *Heinrich* (1949), E.G. 176
Nine Lombardy poplars were planted by the defendant in 1939 on his own land but some 8 feet or so from the wall of the plaintiff's house. During the eight succeeding years the roots spread on to the plaintiff's land and grew under the foundations of his house absorbing water from the surrounding clay soil, with the result that the wall subsided and cracked.

Although a bomb had fallen in 1944 about a mile from the property, evidence showed that the damage was in fact caused by the roots of the poplars. The plaintiff was accordingly awarded damages amounting to some £377.

(ix) *Daisley* v. *B. S. Hall & Co.* (1972), *Estates Gazette*, 3 March 1973
A claim was made by the plaintiff against a firm of chartered surveyors in respect of their alleged negligence. This arose over the survey of a property which the plaintiff wished to buy and which he subsequently bought. In their report on the property, the surveyors made no mention of a row of poplars growing on the boundary of the garden or that the soil was clay over chalk. Cracks developed in the house and since it was held that a surveyor should have drawn attention to both the poplars and the type of soil, damages were awarded to the plaintiff with costs.

(x) *Davey* v. *Harrow Urban District Council* (1957), 2 A.E.R. 305; 2 W.L.R. 941
The plaintiff owned a house which had been built in 1938 on a site which adjoined a piece of land belonging to the defendants. The roots of elm and ash trees alleged to be growing on the defendant's land damaged the plaintiff's house. He therefore brought an action for damages to his property, for negligence in felling some of the trees in 1950 and failing to kill their roots and for an injunction to restrain

the Council from continuing to damage his property by allowing the roots to grow.

The land on which the trees were growing was considered by the plaintiff to be the property of the Council, since they had exercised acts of ownership over it. These included digging a trench on it in order to lay water pipes, and the felling of some of the trees in 1950. The defendants admitted owning land adjacent to part of the plaintiff's property, but denied owning the land on which the offending trees grew. Further, that the trees were growing naturally (as opposed to being planted), had grown there for about 200 years and that they were entitled by prescription to the free growth of roots.

It was held that the trees were not growing on the defendant's land and the fact that some had been felled by the Council did not alter the ownership of the land. Judgment was accordingly given in favour of the defendants.

The plaintiff appealed. In the meantime further information had been obtained that correspondence had passed between the Council and the builders of the plaintiff's house. From this it was clear that the trees were the property of the Council but this had not been disclosed at the original hearing. The appeal was therefore allowed.

Two points of interest arose out of this case. Firstly, that there is no difference in such cases between trees which have been planted or are self-sown. Secondly, that where a boundary hedge is shown on an Ordnance Survey map by means of a line, such line indicates the centre of the hedge.

(xi) *Davis* v. *Artizans Estates, Ltd* (1953), 1953 E.G. 190

The roots of some Lombardy poplars growing on property, of which the defendants were the leaseholders, damaged the main walls, ceilings and drains of the plaintiff's house. The defence was that the defendants did not acquire their property until March 1949 and had not previously been either the owners or the occupiers.

It was held that the defendants were liable for damages to the plaintiff's property but only for that caused after 14 March 1949 when they acquired their leasehold interest. In passing judgment, the Judge held that a landlord could be held liable for damage which was caused by the roots of trees growing on his land when he retained control of the trees.

(xii) *Edge* v. *Briggs* (1960), *Estates Gazette*, 29 April 1961

This case was an appeal by Mr Briggs from a judgment at the County Court which granted Mr Edge an injunction restraining Mr Briggs from causing or allowing the roots of a lime tree to encroach on to Mr Briggs's premises and which granted him some £32 damages.

The appellant (Mr Briggs) was tenant of a house, land and buildings which adjoined the property occupied by the respondent (Mr Edge). Near the buildings, part of which were used as a dairy, there was a lime tree over 60 ft high. The roots of this tree had entered the respondent's garden and caused damage which was agreed at some £32. The appellant was asked to remove the tree but was not prepared to do this. He was, however, prepared to dig up or cut off any encroaching roots. He claimed, however, that he was unable to carry out this work owing to the untidy state of the respondent's garden. The respondent did not clear his garden but instead proposed that an independent surveyor should examine the site and report as to what should be done.

To this the appellant agreed but said that he would not undertake to be bound by the surveyor's decision, and as a result proceedings were started in the County Court. There the judge considered that the removal of the tree as recommended by the surveyor was too drastic, but ordered that the nuisance should be abated by the appellant, probably by cutting such roots which were the cause of the trouble.

The defence submitted that as the appellant (Mr Briggs) was only the tenant, he was under no liability to the respondent (Mr Edge), who was also a tenant under the same landlord. The question was whether the appellant was the right person against whom the action should be brought. It was held that a lessee or tenant of land was liable for nuisance if that nuisance was caused by something on the land which was let to him. The appeal was therefore dismissed.

(xiii) *Jennings* v. *Taverner* (1955), 2 A.E.R. 769 and *Estates Gazette*, 21 May 1955

The plaintiff, as administratrix of her husband's estate, brought an action against a builder in respect of a bungalow which he had erected and which her husband had bought. The bungalow was completed in May 1950 and the plaintiff and her husband occupied it in the same month. By June 1950 a crack which had been noticed by Mr and Mrs Jennings during the erection of the bungalow, and which the defendant had rectified, had reappeared. Subsequently numerous cracks began to develop and although the defendant was informed about them, he did not inspect the property until the middle of 1951. By then the cracks in the kitchen were so large that daylight could be seen through them.

On behalf of the plaintiff, it was alleged that the foundations of the bungalow were not deep enough and had been laid where roots of poplars were present. These trees were growing between 30 and 35 feet away from the house in a cemetery and were clearly visible. It was submitted that a competent builder would or should have known of the danger from their roots. Furthermore that when buying a house which was in the course of construction there were certain implied warranties that the house should be properly built. The trees had since been removed and no further damage had occurred to the premises.

For the defendant, it was submitted that because of ignorance which was common in the building trade at that time, the bungalow had been erected in certain circumstances in which it became defective owing to the presence and effect of the trees. Further that there was no evidence that the dwelling had not been properly built, and that the defendant's liability was limited as at the date of completion.

It was held that the defendant should have known that poplar trees were likely to cause damage to buildings, in view of the literature which had been published warning builders of this danger. Further, when the bungalow was handed over to the plaintiff's husband it had not been built in a workmanlike manner and was not fit for human habitation, since the defendant had failed to prevent the settlement of the walls. Also that there had been an implied breach of warranty. Judgment was therefore given for the plaintiff.

(xiv) *King and Another* v. *Taylor and Others* (1976), *Estates Gazette*, 24 April 1976

Cracks appeared in various parts of the plaintiff's property — a boundary wall, the garage, paths and drains. The plaintiffs claimed damages from the defendants who were the adjoining owners and on whose property were several lime and elm trees (the soil was London clay). They also applied for an injunction to prevent further root encroachment. The plaintiffs were awarded damages with costs, and an injunction was granted against the defendants restraining them from allowing roots of trees on their land to enter the plaintiff's property.

(xv) *Lemos* v. *Kennedy Leigh Developments Ltd* (1960), 105 S.J. 178 and *Estates Gazette*, 4 June 1960 and 11 March 1961

The plaintiffs sought an injunction to restrain the defendants from allowing the roots of trees, growing on land belonging to them, to encroach on their adjoining property. They also claimed damages.

The facts of the case were that the plaintiffs built a house on their land which was some 8 feet from the boundary of the defendants' land, the house being completed in June 1958. In November 1957 the defendants planted a row of twenty Lombardy poplar 7 feet within their boundary and 15 feet from the plaintiffs' house. Roots had been found some 5 feet on the plaintiff's side of the boundary. The plaintiffs considered that these tree roots formed a serious danger to their house and its drains on account of the closeness of the trees and the heavy clay soil in which they were growing. Although the plaintiffs complained, the defendants stated that the trees had not caused and would not cause any damage, and said that they would take immediate action to prevent any damage should it become necessary. Furthermore the defendants dug two trenches near the boundary in order to see whether any roots were developing.

Although it was admitted that no damage to the plaintiffs' property had occurred, a writ was issued on 7 July 1959. In recognizing the risk of damage by the roots of the trees concerned, the defendants offered to remove the trees, to inhibit further root growth and to dig an adequate trench and sever any roots which were revealed. This action the judge considered would remove all reasonable likelihood of damage to the plaintiff's property. The action was accordingly dismissed with costs.

An appeal against the award of costs was dismissed.

(xvi) *McCombe* v. *Read and Another* (1955), 2 Q.B. 429 and *Estates Gazette*, 14 May 1955

The case concerns an action brought by the plaintiff for damages for alleged trespass, nuisance and negligence on account of the roots of poplar trees on the defendant's land encroaching on the plaintiff's property. An injunction to restrain the defendant from continuing the nuisance and trespass was also sought.

When the plaintiff's house was built in 1912, the then owner of the defendant's house planted a row of Lombardy poplars as a screen. The plaintiff complained in 1955 that the roots of these trees had damaged the foundations of his house and caused considerable settlement. Witnesses were called by both sides, but it was held that the subsidence had been caused by the poplar roots and judgment was given for the plaintiff. An injunction was accordingly granted to restrain the defendant from permitting the roots of any tree growing on his land to encroach on the plaintiff's property so as to cause a nuisance. The matter of damages was referred to the Official Referee.

(xvii) *Masters* v. *Brent London Borough* (1977), *Estates Gazette*, 6 May 1978

A claim was made by the plaintiff in respect of damages caused to his property by a lime tree. This had been planted in the pavement outside his house by the defendants or their predecessors in title. He was awarded damages.

(xviii) *Mayer* v. *Deptford and Lewisham Borough Councils* (1959), *Estates Gazette*, 30 May 1959

This case is worthy of special note because it concerns alleged damage to the foundations of a house by the roots of a plane tree and not a poplar, which species figures so frequently in such cases.

The plaintiff brought an action for damages alleged to be caused by a plane growing on the roadside outside her house. Cracks and subsidence had occurred to the premises, but it was alleged by the defendants that these were due to the house having inadequate foundations and that water had escaped from a cracked gulley (caused by the movement of the house) and this had contributed to the settlement. Furthermore the tree was a poor specimen with weak roots.

It was held that although the roots of the plane may have hastened the settlement, they had not caused it. Judgment was therefore given for the defendants.

(xix) *Mills* v. *Smith, Sinclair Third Party* (1963), 3 W.L.R. 367; 2 A.E.R 1078

This case may be divided into two parts: the first dealing with the question of damage to a property by the roots of an oak tree and the second as to whether the owner of the tree was covered by his insurance policy in respect of the claim against him for damages.

The plaintiff alleged that the defendant had allowed the roots of an oak, 150 years old, to encroach on his property and damage the foundations and walls of his house. The soil was London clay, and it was held that the roots of the tree in question were responsible for the damage to the plaintiff's house, and judgment was given accordingly.

The second part of the case was concerned with the defendant's claim that the third party, as one of his insurers, was bound to pay to the defendant any sum which he might have to pay as a result of the judgment. The policy was a Lloyd's Householder's Comprehensive Policy and contained a section which stated that there was a liability to pay 'all sums for which the assured, as occupier, may be legally liable, in respect of claims made by any persons for damage caused to property by accident'. The insurers claimed that the natural growth of a tree and its roots could not be regarded as an accident.

However, it was held that a moment of time would come when the movement of the roots overstepped the safety limit and started a crack in the building and that this would constitute an accident within the context of the policy. Consequently the insurers must indemnify any sums which the plaintiff had to pay and judgment was given against the third party.

(xx) *Murray and Others* v. *Hutchinson* (1955), *Estates Gazette*, 22 October 1955

This is an interesting case since it covers the question of apportionment of damages between damage caused by tree roots and that caused by other factors.

The plaintiffs alleged damage by the roots of three black Italian poplars to the adjoining property of which they were lessees. The house was built on a clay sub-soil, there had been two dry summers in 1947 and 1949, and bombs had fallen in the area during the war. The premises were in such a serious condition that the local authority issued a demolition order and the house was demolished in 1951.

The case for the defendant was that the condition of the premises was due to faulty construction, building on a clay sub-soil which had been seriously affected by two dry summers, the effect of wartime bombing, failure to repair structural defects, and the roots of a plane tree planted beside the roadway and other trees not on the property of which the defendant was owner.

It was held that although the damage to the premises was partly due to the roots of the poplars, there were other causes as well. These were settlement, which was common to the area, dry summers and the effect of wartime bombing. The damage due to the tree roots was apportioned at 25 per cent of the whole and judgment given accordingly.

(xxi) *Niklaus* v. *Moont* (1950), E.G. 174
This case concerns yet another claim for damages in respect of the roots of poplars. There were on the defendant's land five or six poplar trees growing near the boundary of the plaintiff's land. It was alleged that these roots had encroached on to the plaintiff's land and had damaged certain drains and also the wall of the plaintiff's house. Judgment was given for the plaintiff, who was awarded damages.

(xxii) *Pettifer* v. *Cheshunt Urban District Council* (1970), *Estates Gazette*, 19 December 1970
This case dealt with a claim for damages on account of nuisance by encroachment of the roots of two elm trees. The defendants admitted liability, and the point of issue related only to damages.

The plaintiff lived in a semi-detached bungalow, and on the adjoining land, which belonged to the defendants, two large elms were growing. It was alleged that the roots of these trees extracted water from the clay soil and caused the subsidence of the plaintiff's house. The Council admitted that this was a nuisance for which they were liable but they did not agree the extent of the damage or with the steps which it was alleged should be taken to deal with the matter. The plaintiff's surveyor had recommended underpinning, but this the defendants considered was unnecessary. Cracks had developed in the bungalow to such a degree that the plaintiff claimed that he could see buses passing in the road outside. The trees had been felled by the defendants and the roots poisoned.

Damages were assessed by the learned judge in some detail and for particulars of these, reference should be made to the original report. The total awarded to the plaintiff was £1,500.

(xxiii) *Rigby and Another* v. *Sun Alliance & London Insurance Ltd* (1979), *Estates Gazette*, 3 November 1979
This somewhat involved case arose out of an action for the alleged damage caused to a bungalow by the roots of several oak trees. However, the case is primarily concerned with the responsibility of two insurance companies for claims arising out of the alleged damage, and with the definition of 'nuisance' under the policy.

(xxiv) *Russell and Another* v. *London Borough of Barnet* (1984), *Estates Gazette*, 25 August 1984
The plaintiffs sued the defendants in respect of nuisance and negligence for damage to their property, alleged to have been caused by the roots of two oak trees which were growing in the pavement outside their house. The soil was London clay. The defendants, as highway authority, denied ownership of the trees or that they were responsible for them. This was on the grounds that the owners of land adjoining the highway are also owners of the sub-soil out of which the trees were growing. Judgment was given for the plaintiffs.

(xxv) *Solloway* v. *Hampshire County Council* (1981), *Estates Gazette*, 30 May 1981
This was an appeal by the County Council (which was also the highway authority) from a previous decision which concerned a claim against the Council for damage to the plaintiff's house by the roots of a horse chestnut. In that case the judge had decided in favour of the plaintiff.
Although the soil was plateau gravel, small scattered pockets of clay occurred in it but these were not large enough to be shown on the geological map of the area. As it happened, the rear and part of the side of the house, stood on gravel while the front was on one of the clay pockets. The outcome of the appeal depended on whether, in these circumstances, the highway authority could reasonably have forseen the risk of subsidence brought about by the presence of the roots. It was held that the authority could not have done so, and the appeal was allowed.

(xxvi) *Wallace* v. *Clayton* (1961), *Estates Gazette*, 3 March 1962
This case concerned four poplar trees growing close to the boundary between two properties. The plaintiff, who was the lessee of a flat, brought an action against the defendant

for damages alleged to have been caused in two ways. Firstly by the roots, which were said to have caused cracks in a bedroom and bathroom, and secondly through two of the trees exerting physical pressure on a wall and glass roof of a wash-down. The soil was derived from the Claygate Beds, and although alleged by the plaintiff to be a clay soil, much evidence was produced to contest this view and to show that this soil contained a substantial proportion of sand, the purpose of such evidence being to show that poplar roots did not have the same effect on a soil containing sand as they did on a heavy clay.

The plaintiff was awarded damages and an injunction similar to that granted in *McCombe* v. *Read and Another* (1955), of which case a short account is given earlier in this section.

(xxvii) *Watson and Roberts* v. *Smith and Wakeham* (1956), *Estates Gazette*, 14 July 1956
This case concerned some poplar trees which were growing in the defendant's gardens, the roots of which had caused damage to the plaintiff's houses. The parties came to terms under which each defendant paid each plaintiff £275 and costs.

Summary of species
The following is an abstract of the species mentioned in the 27 cases which are outlined above:

Poplars	13	Horse chestnut	1
Oak	4	Ash and elm	1
Elm	2	Poplar and elm	1
Lime	2	Lime and elm	1
Plane	1	Various	1

(c) **Poisonous trees**
  (i) *Cheater* v. *Cater* (1918), 1 K.B. 247, C.A.
The plaintiff was the tenant of the defendant, and a field which was occupied by the defendant was separated from a field which he had let to the plaintiff by a yew hedge. The hedge overhung the plaintiff's land by about 3 feet, and a horse which belonged to the plaintiff ate some of the yew hedge and consequently died. It was held that the defendant, who was the landlord, was not liable, in accordance with the judgment in the case of *Erskine* v. *Adeane* (1873), 8 Ch. App. 756, which is reported below. The plaintiff appealed, but this was dismissed.

(ii) *Crowhurst* v. *Amersham Burial Ground* (1878), 4 Ex. D. 5
The defendants planted a yew tree on their own land and about
4 feet from the boundary fence. In course of time as the tree
grew, some of the branches projected through and over the
boundary fence. Parts of the tree consequently became
accessible to the plaintiff's horse, which was grazing in the
adjoining field.

As a result of this the horse died from yew poisoning and the
plaintiff brought an action for damages. It was held that the
defendants had allowed the tree to grow over the boundary and
that they were consequently responsible. This follows the
ruling given in *Rylands* v. *Fletcher* (1868), 3 H.L. 330, namely
that if a person brings on to his land something which is
dangerous or harmful and he allows it to escape, he will be held
liable for any injury which may result.

(iii) *Erskine* v. *Adeane* (1873), 8 Ch. App. 756
The defendant let a farm, which he owned, to the plaintiff and
under the terms of the letting reserved to himself a plantation
which contained some yew trees. Some of the plaintiff's farm
stock subsequently ate some of the yew foliage and died as a
result. The defendant would not agree to erect a fence around
the plantation before the stock were poisoned.

It was held that the defendant, as landlord, was not liable,
since the yew trees were quite apparent when the plaintiff
agreed to rent the farm. Furthermore he could have required
the defendant, as landlord, to maintain a stockproof fence
around the plantation as a covenant when taking the farm. As it
was, he had taken the farm with his eyes open and without
asking for such a covenant.

(iv) *Lawrence* v. *Jenkins* (1873), 8 Q.B. 274
The defendant sold some foliage from trees growing in his
wood and the purchaser negligently broke down the fence
between the defendant's wood and the plaintiff's land. Some
cows which were the property of the plaintiff gained access to
the wood through the damaged fence and were found next
morning beside a yew tree, and died later that morning. It was
the defendant's obligation to keep the fence in repair.

Although the buyer of the foliage had broken down the fence
and although the defendant did not know that the fence had
been damaged, it was held nevertheless that the defendant was
liable, and damages were awarded accordingly.

(v) *Ponting* v. *Noakes* (1894), 2 Q.B. 281

A yew tree was growing on the defendant's land but its branches did not reach or extend over the boundary between land belonging to the defendant and that belonging to the plaintiff. A horse belonging to the plaintiff reached over the boundary fence and ate some of the yew foliage, with fatal results.

Judgment was given in favour of the defendant, since it was held that he was under no obligation to prevent the plaintiff's horse from trespassing by reaching over the boundary fence and eating the yew's foliage.

(vi) *Wilson* v. *Newbury* (1871), 7 Q.B. 31

The defendant owned a number of yew trees growing on the land in his occupation. He was aware that the clippings of yew trees were poisonous to horses and cattle.

It was alleged that the defendant did not take proper care to prevent clippings from the yew trees from being placed on the plaintiff's land. As a result these were eaten by a horse belonging to the plaintiff, which consequently died.

It was held that the defendant was not responsible, since there was no evidence that he had in fact clipped the yews or that he knew that the trees had been clipped or that he had placed the clippings on the plaintiff's land.

(d) *Dangerous trees*

(i) *Brown* v. *Harrison* (1947), 177 L.T. 281 and 1947 E.G. 158

As the plaintiff was passing along a road, a horse chestnut which was standing some 18 feet from the highway fell on him. It was an old tree, the top branches of which were dead, and although there was a high wind blowing at the time it could not be considered exceptional.

The defendants were held to be responsible by the County Court, and their subsequent appeal was dismissed since the condition of the tree was such that it was evident that it was dangerous.

(ii) *Bruce* v. *Caulfield* (1918), 34 T.L.R. 204, C.A.

The top of a poplar was blown on to the plaintiff's stables during a very severe gale and caused considerable damage. The plaintiff claimed damages on the grounds of trespass, nuisance and negligence, and was awarded them by the County Court Judge.

On appeal, however, it was held that there was no evidence

of negligence, that the question of nuisance had no bearing on the case, and that there was no evidence to show that the defendant had caused the tree to fall. The appeal was therefore allowed.

(iii) *Caminer* v. *Northern and London Investment Trust Ltd* (1951), A.C. 88 H.L.
A large elm about 130 years old fell on to the plaintiffs' car as they were driving past, and caused injury to both of them. After it had fallen, it was seen that the tree was suffering from heart rot and that some of its roots were badly affected. At the first hearing of the case judgment was given for the plaintiffs. On appeal, however, the decision was reversed, since it was held that the heart rot was not apparent and a careful inspection would not have revealed it. There was no evidence to show that the defendants had been negligent or had failed in their duty.

(iv) *Cunliffe* v. *Bankes* (1945), 1 A.E.R. 459 and 1945 E.G. 98
An elm about 50 years old, which was subsequently found to be suffering from honey fungus, fell across a public road. The plaintiff's husband, who was riding his motor cycle along the road, collided with the tree and later died from the injuries which he sustained. The plaintiff brought an action against the defendant based on negligence and nuisance under the Fatal Accidents Act 1846, and the Law Reform (Miscellaneous Provisions) Act 1934. It was shown that the defendant had taken all reasonable steps to discover what the condition of the tree was and that he could not have known that it was likely to fall.

It was held that the defendant had not been negligent and that he was not liable for nuisance. A person is not liable for nuisance constituted by the condition of his property unless he causes it, or by neglect of some duty he allows it to occur, or, if it should arise without his own act or default, he fails to remedy it within a reasonable time after he had become, or ought to have become, aware of it.

(v) *Hudson* v. *Bray* (1917), 1 K.B. 520
When driving at night the plaintiff struck an elm tree which had been blown down on the previous day. He was awarded damages on the grounds that the defendant having been informed of the position had failed to place warning lights near it.

On appeal the defendant was held not to be liable for the fall of the tree, although he would have been liable for failing to remove it, if he had been served with notice to do so, under the Highways Act 1835. In fact he had only been given notice to illuminate it.

Reference should also be made to the case of *Williams* v. *Devon County Council* (1966), *Estates Gazette*, 10 December 1966.

(vi) *Kent* v. *Marquis of Bristol* (1940), *Quarterly Journal of Forestry*, January 1947

An elm blew down during a strong wind and fell on a passing milk van, killing the driver. On examination, a hole was found in the tree 4 in. × 8 in. and 12 to 15 in. deep. It was held that if the tree had been inspected the hole would have been seen and it would have indicated the probable existence of decay. Apparently the tree was not inspected and no action had been taken to deal with the matter. Consequently the plaintiff was awarded substantial damages.

(vii) *Knight* v. *Hext and Others* (1979), *Estates Gazette*, 22 March 1980

This was an appeal against a decision given in favour of Mr F. C. Knight in respect of a beech growing on property owned by Mrs Hext which fell on to a barn belonging to the plaintiff. The relevant facts were that Mrs Hext had moved into the property only on 12 December 1972 and the tree had fallen on 27 December. On 16 December, she saw Mr Knight cutting up a tree in the park which was then his property and she spoke to him. During this conversation it was alleged that Mrs Hext remarked that some trees should be felled, and pointed to a clump in which stood the tree which later fell on the barn.

The question was whether there was any urgency about the matter and whether Mrs Hext should have inspected the tree between 16 and 27 December, bearing in mind that there were only $5\frac{1}{2}$ working days when allowance was made for Christmas. The Court agreed that she did not have reasonable time to do this, and her appeal was allowed.

(viii) *Lambourne* v. *London Brick Co.* (1950), *Estates Gazette*, Vol. 156, p. 146.

Some elm trees were blown across a road, and the plaintiff's husband crashed into them and was killed.

Since there was nothing to show that the trees were dangerous, it was held that the owners of the trees were not liable.

(ix) *Lane* v. *Tredegar Estate Trustees* (1954), 1954 E.G. 216.

A branch from a horse chestnut belonging to the defendants fell into the road beside which it was growing, and a motor van driven by the plaintiff crashed into it.

During the hearing of the case there was a considerable amount of conflicting evidence between the witnesses called by the opposing sides. The branch which had fallen was apparently rotten at the centre. Other branches which had been removed at an earlier date were also rotten at the centre, and it was held that this fact should have aroused the suspicions of the defendants and that they should have taken action, which they did not.

It was further held that the branch fell on account of the defendants failing to exercise reasonable care and that they were therefore responsible. Damages were awarded to the plaintiff.

(x) *Mackie* v. *Western District Committee of Dumbartonshire County Council* (1927), *Justice of the Peace and Local Government Review Reports*, 1927, vol. 91, p. 158

An elm growing on private property, but in close proximity to the highway, fell on to a passing motor charabanc killing some of the passengers and injuring the appellant. He brought an action against the highway authority, on the grounds that they were to blame in not having inspected the road to see if it was safe and, further, in not causing a dangerous tree to be removed. At the point at which the accident occurred, the road had been widened by the County Council twelve months before the accident and a quantity of soil had been removed from the base of the tree and some of its roots had been cut. It was shown that as a result it was obvious that the tree would probably fall.

This Scottish case involved a point of procedure as to whether the pleadings disclosed that there was a cause of action or not. It was held that the facts did disclose a relevant cause of action.

(xi) *Noble* v. *Harrison* (1926), 2 K.B. 332

The branch of a beech about 80 years old, which was overhanging the highway, fell on a calm day on to a passing bus which belonged to the plaintiff. The County Court Judge held that, although the defendant did not know that the branch was dangerous or that it was in fact fractured and that this fracture would not have been revealed by a reasonably careful inspection, he was nevertheless liable to the principles of *Rylands* v. *Fletcher* (1868), L.R. 3, H.L. 330. This case is referred to in Section 2 of this chapter.

On appeal, however, the County Court decision was reversed as it was held that the case of *Rylands* v. *Fletcher* did *not* apply. Furthermore although the branch overhung the highway, it did not constitute a nuisance because it did not obstruct the highway until it fell. Although the branch proved to be a danger, the defendant was not liable since he had not created the danger and had no knowledge that it existed. Judgment was given in favour of the defendant.

(xii) *Quinn* v. *Scott and Another* (1964), 1 W.L.R. 1004; 1964 E.G. 441; and *Estates Gazette*, 27 March 1965 (Vol. 193)
A 200-year-old beech growing on private property fell on to the adjoining highway in front of fast-moving traffic. As a result a car crashed through some of the branches and hit a small bus driven by the plaintiff.

It was stated that the tree appeared unhealthy since its foliage was light and there were signs of die-back.

It was held that a prudent owner of the tree would have felled it in view of the indications of unsoundness which the tree displayed. The owners of the tree were consequently held to be liable.

(xiii) *Sheen* v. *Arden* (1945), *Journal of the Land Agents' Society*, January 1946
Shortly before midnight a tree was blown down across a highway and a few hours later was struck by the plaintiff's husband, who died shortly afterwards. It was shown that the tree which fell was one of a clump and, although previously sheltered by its fellows, had become exposed to the full force of the wind after adjoining trees on the windward side had been removed.

It was held that the owner of the trees was responsible since, having removed the trees on the exposed side, it was an obvious and elementary precaution to see if the remaining trees were able to withstand the new stresses to which they would be subjected.

(xiv) *Shirwell* v. *Hackwood Estates Ltd* (1938), 2 K.B. 577, C.A.
In January 1936 the owner of an estate let part of it to a tenant while he himself remained in occupation of the adjoining land. In February the owner sold the estate subject to the tenancy, and the new owner continued to occupy the adjoining land. On this land and near the boundary of the land which was let was a beech tree which was dead from a height of 16 feet above the ground to the top. In May a branch fell and killed a farm worker who was the tenant's servant and working on the tenant's land.

At the beginning of the tenancy there were many dead and dying trees on the estate, some of which had been felled, and advice was being sought as to the remainder.

The tenant paid compensation to his employee's widow under the Workmen's Compensation Acts of 1925–34 and then brought a case against the owner of the tree for indemnity. However he lost his case but appealed, and the Court of Appeal ruled that there was no case against the owner for the following reasons.

Firstly, the tenant had taken the land with the dead trees on it and had not objected. Secondly, even if the owner was under a duty to the tenant to protect him from falling branches (and the Court considered that he was not) then the owner had taken reasonable care to discharge such duty. And thirdly, the owner was under no higher duty to the tenant's servants than he was to the tenant.

(xv) *Stillwell* v. *New Windsor Corporation* (1932), 147 L.T. 306
The land belonging to the plaintiff adjoined a Highway which consisted of a carriage-way and a footpath between the carriage-way and the plaintiff's land. Between the carriage-way and footpath some trees were growing. The Corporation served notice on the plaintiff requiring her to remove the trees as they were considered to be a danger and obstruction to members of the public when they were using the highway.

The plaintiff refused to comply with the notice and the Corporation therefore cut down three of the trees. The plaintiff then issued a writ against the Corporation, firstly for an injunction to prevent them felling any more trees and secondly for damage in respect of the trees which had already been felled.

However, it was held that the trees were an obstruction

since they had been planted in a situation over which, as an ancient highway, the public had rights of passage. It was further held that if the Corporation considered that the obstruction was a serious one, it was their duty to remove it. Judgment was therefore given in favour of the Corporation.

(xvi) *Thomas* v. *Miller* (1970), *Western Morning News*, 3 February 1970

The plaintiff, who was on holiday, parked his car in a lane which he admitted he knew was private property. When he returned to his car he found that the branch of an elm tree had fallen on it and that the car was a total loss. The lane and the tree concerned belonged to the defendant.

The defendant stated that he was at the property only from time to time at Easter and during the summer, but it was his habit to inspect the property at least twice a year to see the trees and fences. He had never tried to stop holiday makers parking their cars in the lane and he considered that it was almost impossible to do so, but on three or four occasions during the season he would speak to those concerned and remind them that the property was private and that they parked their cars on it at their own risk.

Two expert witnesses were called regarding the condition of the tree in question. The first witness stated that he had inspected the tree about a year after it had fallen and found it to be very decayed at the base. The second witness said that he had found the tree alive and the wood sound, only the stump of a former limb was decayed.

The County Court judge found in favour of the plaintiff and in doing so said that the plaintiff and others who parked in the lane were visitors and not trespassers. He further considered that it was clear that the defendant had some suspicions that the trees adjoining the lane were unsafe.

(xvii) *White* v. *Carruthers* (1958), *Estates Gazette*, 16 August 1958

A caravan belonging to the plaintiff was kept in a caravan park which was the property of the defendant. In April 1957 an elm tree fell on to the caravan and the plaintiff claimed in respect of the damage to it, for the contents and for the cost of alternative accommodation while the caravan was unusable.

The plaintiff alleged that the damage was caused through the defendant's negligence, which was denied. She, the plaintiff, had made a series of complaints regarding the condition of the trees near her caravan to the caravan site manager, but these had not been passed on to the defendant.

In giving judgment in favour of the defendant the County Court Judge said that the defendant had 'kept an eye on the trees' and also employed a timber feller to go round at least twice a year to inspect the trees. He considered that the defendant discharged his duties as a landowner with reasonable care. The judge said that the tree which had fallen was infected with heart rot but that there was no external evidence of it. The plaintiff's claim was accordingly dismissed.

Note. These last two cases, *Thomas* v. *Miller* and *White* v. *Carruthers*, are of particular interest since they were heard before the same judge at the same County Court.

(xviii) *Williams* v. *Devon County Council* (1966), *Estates Gazette*, 10 December 1966

This case was an appeal by the Devon County Council against a decision of the Devon Quarter Sessions which had allowed an appeal by Sir Robert Williams, Bt., from a finding of the local magistrates that he was liable to pay £30 10s. to the highway authority in respect of the cost of removing a tree growing on his land and which had fallen across the highway.

The facts were that in January 1965, during a night of high winds, an elm growing on Sir Robert's land was blown down and fell across the highway. Owing to the urgency of the matter the Devon County Council sent some men to remove it. It was proved that the tree was sound and had been under continuous supervision by the owner and his employees. It was thus inferred that the owner had taken reasonable care to ensure that the tree did not fall on to the highway. The County Council as highway authority held that it was necessary for the owner to prove that he had not only taken reasonable care to see that the tree did not fall, but had also taken reasonable care to ensure that the tree, after it had fallen, was removed from the highway.

In giving judgment, Lord Parker CJ said that the Highways (Miscellaneous Provisions) Act 1961 should be read with the Highways Act 1957. Section 9 of the 1961 Act provided that the highway authority could recover from the owner of the thing which caused the obstruction, such expenses as were reasonably incurred in its removal. However, no expenses could be recovered from someone who proved that he had taken reasonable care to ensure that the thing in question did not cause or contribute to the obstruction. The appeal was accordingly dismissed with costs.

Summary of species

The following species were involved in the 18 cases referred to above.

Elm	9	Poplar	1
Beech	4	Unspecified	2
Horse chestnut	2		

(e) Tree preservation orders

(i) *Attorney General* v. *Melville Construction Co. Ltd* (1968), *The Times*, 21 August 1968

This case dealt with the granting of an injunction, until the trial of the action or further order, restraining the defendants from cutting down or wilfully destroying any trees protected by a tree preservation order. The defendants were developing some land which was the subject of the order concerned, and in the course of the work some trees had been damaged and two destroyed. Once a tree had been destroyed it could not be replaced and it would not be right that the defendants should be free to injure or fell trees during the time when they were being prosecuted for breach of the preservation order. An injunction was therefore granted.

(ii) *Barnet London Borough Council* v. *Eastern Electricity Board and Others* (1973), *Estates Gazette*, 21 April 1973

This was an appeal by the Borough of Barnet against an order made by the Highgate justices which had dismissed information brought by the Council against the Eastern Electricity Board and its contractors. It arose over the re-laying of cables which were close to six horse chestnut trees, the subject of a tree preservation order. In carrying out the work, the roots were cut or damaged, and it was alleged that the lives of the trees had been shortened and that they were unstable and possibly dangerous. Subsequently four were felled by the Council and two were lopped.

It was maintained by the appellants that the severing of the roots constituted 'wilful dstruction' of the trees contrary to the tree preservation order. This view, which was opposed by the respondents, had been upheld by the magistrates who decided that the Board had not contravened the preservation order.

However, the Court was of the opinion that a tree covered by such an order was destroyed when something was done to it which made it no longer an amenity or worth preserving. The case was sent back to magistrates so that further evidence could be called.

(iii) *Bell* v. *Canterbury City Council* (1988), *Estates Gazette*, 4 June 1988

In this case the Council appealed against a decision of the Lands Tribunal, the facts of the case being as follows. Mr P. J. Bell, the owner of the land, began to clear the area with a view to converting it from woodland to agriculture. However, a tree preservation order was placed on the area, and when Mr Bell applied for permission to deal with a further 39 acres, it was refused.

He was subsequently awarded compensation by the Lands Tribunal amounting to £38,851, which was based on the difference between the value of the land as woodland and as agricultural land. The Council appealed against this decision, claiming that compensation should be limited to the commercial value of the trees which could not be felled under the preservation order. The appeal was dismissed.

(iv) *Bellcross Co. Ltd.* v. *Mid Bedfordshire District Council* (1987), *Estates Gazette*, 16 April 1988

This somewhat involved case arose from an application by a development company – Bellcross Company Ltd – to quash a tree preservation order made by the Mid Bedfordshire District Council. Very briefly, the main points are as follows.

The company was granted planning permission for the residential development of the area, and some six months later the Council served a preservation order on it. This was in respect of a single mature yew tree growing in the garden of a farm house which was to be demolished and the site redeveloped. The Company objected to this, and about three weeks later wrote stating that they had amended their application and enclosed a tree surgeon's report. This suggested that it would be dangerous to retain the tree, and for this reason the Company objected to the order. However, the county forestry officer was of the opinion that provided there was no development closer than 10 metres to the tree, there was no reason to remove it. The application was dismissed with costs.

(v) *Bullock* v. *Secretary of State for the Environment and Another* (1980), *Estates Gazette*, 21 June 1980

This was an application from Mr J. Bullock to quash a tree preservation order that had been placed on a small wood, half an acre in extent and consisting mainly of ash coppice which he owned. When the applicant applied to build eight houses on the site, he was refused permission on amenity grounds and also because there was other land available for development. A tree preservation order was placed on the wood when it was found that the applicant had felled several trees.

It was held that a tree preservation order could be made in respect of coppice and, in such an event, that the area could still be managed satisfactorily. The application was dismissed with costs, in favour of the Secretary of State.

(vi) *Edgeborough Building Co. Ltd* v. *Woking Urban District Council* (1966), *Estates Gazette*, 14 May 1966

This was an appeal by the Edgeborough Building Co. against a conviction for an infringement of a tree preservation order which had been made by the Urban District Council in 1963. The appellants had obtained permission from the respondents to build a house on a plot of land on which was growing a horse chestnut tree which was subject to a tree preservation order. The house was erected some 12 to 20 feet from the tree, but when a prospective purchaser said that he was afraid to buy the house because of the risk of damage by the roots of the tree, the appellants felled it without consulting the Council or obtaining their consent or taking advice on the matter.

To succeed in their appeal the company had to show that they had felled the tree either to prevent or abate a nuisance or in order to carry out development for which planning permission had been granted. The submission that the presence of the tree constituted a risk of damage to the property was not accepted by the magistrates at the original hearing. The Court endorsed the magistrates' action and the appeal was dismissed.

(vii) *Maidstone Borough Council* v. *Mortimer* (1980), *Estates Gazette*, 6 December 1980

The Council appealed against the action of the local magistrates in acquitting the respondent who, it was alleged, had contravened a tree preservation order. The respondent who was a tree feller, was given the task of felling a mature oak by the owner of the tree – a Mrs Twydell. The tree was the subject of a preservation order, but Mrs Twydell told the respondent that she had obtained permission to fell it. She was however mistaken and no permission had been given. After Mr Mortimer, the feller, had made one cut, he was informed by the vice-chairman of the parish council that he thought that the tree was covered by a preservation order and asked him to stop. However, acting on what Mrs Twydell had told him, he made another cut and as a result the tree was considered dangerous and had to be felled.

It was held that the fact that the respondent did not know of the preservation order was no defence, and the appeal was allowed.

(f) Covenants relating to trees

(i) *George* v. *Reeves* (1969), *Estates Gazette*, 10 March 1969

This was an appeal by Dr Reeves against a County Court decision ordering him to forfeit a 51 year lease of the wing of a house, on the grounds of breach of covenant, and to pay his landlord, Mrs George, £1 damages upon repayment by her of a premium of £2,500 which he had paid her for the lease in 1963.

That part of the garden which the appellant leased was overgrown, and under his lease he was responsible for attending to it. In the garden were one elm and six Monterey cypresses (*C. macrocarpa*) which he was not allowed to fell without the landlord's consent, such consent not to be unreasonably withheld.

The facts were that Dr Reeves wished to fell the seven trees but Mrs George did not approve. The County Council were consulted and made a tree preservation order. The appellant then ring-barked the elm, which subsequently fell, and also some of the Monterey cypresses. He further interfered with attempts by experts to save the latter.

It was held that the appellant had committed a breach of the covenant of the lease by felling the trees without the written consent of the landlord. Further that the appellant's argument that ring-barking the trees was not felling them could not be accepted since, if after ring-barking a tree died, it was the equivalent of cutting it down. The appeal was consequently dismissed.

(ii) *Molony and Others* v. *Knapton* (1966), *Estates Gazette*, 29 January 1966

This case was an action by the plaintiffs claiming damages and an injunction to restrain the defendant from cutting down trees in his garden which were now or likely to become 15 feet high.

The house belonging to the defendant (Mr Knapton) stood in a large garden in which some 40 to 50 trees were growing and lay to the south-east of the plaintiff's properties. Before the defendant's house was built a number of people, including Mr Knapton's predecessor in title and the plaintiffs, executed a deed in 1958 under which his predecessor covenanted briefly as follows. Firstly that only one house should be erected on the land, and secondly that no tree which reached or should subsequently reach 15 feet in height should be felled or destroyed unless such action was necessary to allow the house to be built.

In 1962 the defendant wished to build an extension of his house and he accordingly prepared a plan which he showed to all concerned and obtained their consent. When the work began in 1963 difficulties arose between Mr Knapton and one of the plaintiffs (Mr Whitehouse) regarding windows and the erection of a garage. Mr Knapton then felled six trees which were dead, only one of them standing on the land which was subject to the covenant. He also did a considerable amount of pruning. Mr Whitehouse formed the opinion that such action was a breach of the covenant and communicated his opinion by shouting to Mr Knapton from his garden gate. Mr Whitehouse and the other plaintiffs then decided to take action and consequently brought the case.

It was held that the defendant had not committed a breach of covenant, and it was admitted that Mr Whitehouse's belief that there had been such a breach was unjustified. The case was dismissed with costs.

(iii) *Richardson* v. *Jay* (1960), *Estates Gazette*, 9 July 1960
The plaintiff claimed first for damages for escape of water from the defendant's garden, which it was alleged had caused the plaintiff's house to subside, and secondly for an injunction restraining the defendant from preventing the plaintiff from cutting down two trees dividing the adjoining gardens of the two properties.

As regards the trees, a dividing fence had been erected between them in 1928 by the first purchaser of the property. When the defendant subsequently replaced it with a more modern fence, the plaintiff alleged a slight encroachment. This led to proceedings in the County Court, and as a result it was agreed that the new fence should follow the exact line of the old one, thus leaving half of each tree in the respective gardens.

The plaintiff now claimed ownership of the whole of the trees, and in support contended that the fence originally had been erected in the wrong place. His claim was based on the Land Registry plan attached to his conveyance of the property.

It was held that the two trees were owned by the parties in common. The plaintiff was entitled to cut the roots on his side of the fence but he had no right to fell the entire trees without the consent of the defendant. The plaintiff was granted a stay of execution pending a possible appeal.

(g) Damage to young plantations
Smith v. *Smale* (1954), *Estates Gazette*, 30 October 1954

In 1946 the plaintiff assigned to the defendant an agreement by which she had let her farm other than the farmhouse. She had also reserved to herself the minerals, gravel, trees and underwood on or under the land, and the tenant undertook to preserve the trees from injury.

In June 1952 she found a gate between one of the fields and a gate tied back and cattle grazing in the wood. Many of the trees had been damaged, some having been eaten off and others trodden down. When she spoke to the defendant he said that he wanted his cattle to eat the grass in the wood.

Judgment was given in favour of the plaintiff and she was awarded damages with costs.

(h) Definitions
Definitions of certain matters concerning trees are set out in the following cases which are not summarized, as in the foregoing cases, owing to lack of space:

 (i) Definition of 'coppice':
    Dashwood v. *Magniac* (1891), 3 Ch. 306.

 (ii) Definition of 'hedgerow timber':
    Layfield v. *Cowper* (1694), 1 Eag. & Y. 591.

(iii) Definition of 'lopping':
    Unwin v. *Hanson* (1891), 2 Q.B. 115.

(iv) Definition of timber:
    (1) According to age and species:
        Honywood v. *Honywood* (1874), 18 Eq. 306.
    (2) According to size:
        Whitty v. *Lord Dillon* (1860), 2 F. & F. 67.
    (3) Alder:
        Goodall v. *Perkins* (1694).

(4) Beech:
> *Abbots* v. *Hicks* (1696), 1 Wood 319.
> *Aubrey* v. *Fisher* (1809), 10 East. 446.
> *Bibye* v. *Huxley* (1824), 1 Eag. & Y. 805.
> *Dashwood* v. *Magniac* (1891), 3 Ch. 306.
> *Layfield* v. *Cowper* (1694), 1 Eag. & Y. 591.
> *R.* v. *Minchinhampton* (1762), 3 Burr. 1309.
> *Walton* v. *Tryon* (1751), 2 Gwill. 827.

(5) Birch:
> *Countess of Cumberland's case* (1610), Moore K.B. 812.
> *Foster & Peacock* v. *Leonard* (1582), 1 Cro. Eliz. 1.

(6) Horse chestnut:
> *R.* v. *Ferrybridge* (Inhabitants) (1823), 1 B. & C. 375.

(7) Hornbeam:
> *Soby* v. *Molyns* (1575), Plowd. 470.
> *Turner* v. *Smith* (1680), 1 Eag. & Y. 526.

(8) Larch:
> *In re Harrison's Trust* (1884), 28 Ch. D. 221.

(9) Spruce:
> *In re Tower's Contract* (1924), W.N. 331.

(10) Walnut:
> *Duke of Chandos* v. *Talbot* (1731), 2 P. Wms. 601.

(11) Willow:
> *Cuffly* v. *Pindar* (1616), Hob. 219.

Abbreviations used in reported law cases
To economize in space the following recognized abbreviations have been used in the law cases reported above. Where there is any doubt or if an accepted abbreviation is not available, the source of the case has been given in full, e.g. *Estates Gazette*, 10 December 1966.

A.C.	Appeal Court
A.E.R.	All England Law Reports
B. & C.	Barnewall and Cresswell's Reports
Burr.	Burrow's Reports
C.A.	Court of Appeal
Ch. D.	Chancery Division Reports
Ch. App.	Chancery Appeal
Cro. Eliz.	Croke, time of Elizabeth
Eag. & Y.	Eagle & Young's Collection of
E. & Y.	Tithe Cases
East	East
E.G.	Estates Gazette Digest of Cases
Eq.	Equity

Ex. Div. ⎫ Ex. D. ⎬	Exchequer Division
F. & F.	Foster & Finlason
Gwill.	Gwillim
H.L.	House of Lords
Hob.	Hobart
K.B.	King's Bench
L.T.	Law Times Reports
Moo. & M.	Moody & Malkin
Moore	Moore
P. Wms.	Peere Williams
Plowd.	Plowden
Q.B.	Queen's Bench
S.J.	Solicitor's Journal
T.L.R.	Times Law Reports
W.L.R.	Weekly Law Reports
Wood	Wood's Tithe Causes

7 *Insurance against damage and injury*

(*a*) Although this section does not come within the strict scope of trees and the law, it is included in this chapter because it is very relevant to the problem of damage and injury which may be caused by trees.

(*b*) Damage by trees can be divided into the following broad classes:

 (i) Damage caused by a tree or part of a tree falling.

(ii) Damage caused by the roots of a tree.

Both of these types of damage may again be considered under two headings:

 (i) Damage to the policy holder's own property.

(ii) Damage to property which is not the property of the policy holder.

(*c*) Damage which is caused to the policy holder's property as a result of a falling tree or branch is usually covered by the 'storm and tempest' section of the policy. However, no claim is likely to be met in respect of dead or rotten trees which fall, except in a storm.

(*d*) Where a tree belonging to a policy holder falls and causes injury or damage to a third party or his property, the tree owner's comprehensive policy usually includes (in the third party section of the policy) cover of up to £100,000 for accidents for third parties. It should be noted that in order to claim compensation a third party must prove that the accident was due to negligence on the part of the tree's owner. For example, it must be shown that he had failed to ascertain if the tree was dangerous or not.

(e) In the case of an extensive property or where there are a great number of trees, it is essential to ensure that the Public Liability Clause or Third Party Section covers all possible risks.

(f) As regards damage which may be caused by tree roots, this matter is seldom if ever covered by a comprehensive policy. Although it might be possible to arrange a special policy to deal with the matter, if there are trees growing on a heavy clay soil close to houses, it is unlikely that any insurance company would undertake the risk.

CHAPTER XVI

Felling Licences

1 *When felling licences are needed*

(*a*) The regulations relating to felling licences and restrictions on felling, are contained in the following statutes and Statutory Instruments (SI):

Forestry Act 1947 as amended by Forestry Act 1979, Section 2 and set out in Schedule 1 of that Act.

SI 791/1979: The Forestry (Felling of Trees) Regulations 1979

1958/1985: The Forestry (Modification of Felling Restrictions) Regulations 1985

1572/1985: The Forestry (Exceptions from Restriction of Felling) (Amendment) Regulations 1985

632/1987: The Forestry (Felling of Trees) (Amendment) Regulations 1987.

(*b*) Section 9(1) of the Forestry Act 1967 states that a felling licence is required in order to fell growing trees except in certain cases which are set out in subsections (2),(3) and (4) of Section 9, as modified by SI 1958 of 1985.

2 *When felling licences are not needed*

(*a*) Section 9 (as amended) also lays down that a felling licence is *not* necessary in certain cases which are set out in subsections (2), (3) and (4) as follows:

(*b*) *Subsection (2)*
   (i) Trees which have a diameter not exceeding 8 cm (3 in.) or coppice and underwood of a diameter of not over 15 cm (6 in.).
  (ii) Fruit trees or trees growing in an orchard, garden, churchyard or public open space.
 (iii) The heavy pruning of trees referred to in the Act as 'lopping and topping' or the cutting and laying or trimming of hedges.

(c) *Subsection (3)*
 (i) Trees which have a diameter of not more than 10 cm (4 in.)
     which are felled in order to improve the growth of other trees.
 (ii) This subsection also laid down the volume of timber that could
     be felled or sold in any one quarter. However, SI 1958 of 1985
     reduced the amounts permitted in the Act as given below.
 (iii) Where the total volume felled in any one quarter does not
     exceed, in the aggregate, 5 cubic metres (138 Hoppus feet).
 (iv) Of these amounts, not more than 2 cubic metres (55 Hoppus
     feet) may be sold, although these amounts can be increased by
     the Forestry Commission.

(d) *Subsection (4)*
 (i) Where felling is carried out so as to prevent a danger or
     nuisance or to abate a nuisance.
 (ii) Where felling is required under an Act of Parliament.
 (iii) Where the trees are felled at the request of an Electricity Board
     on account of their interference with the construction or
     maintenance of an electricity supply line.
 (iv) When felling is needed immediately in order to carry out
     development for which planning permission has been obtained.

(e) *Subsection (5)*
This subsection authorizes the Forestry Commissioners to modify
subsections (2) and (4) in respect of sizes and volumes of trees that
may be felled without a licence.

(f) *Subsections (6)*
 (i) This states that the diameters referred to in subsection (2) shall
     be measured over bark at 1.3 m (5 ft) above ground level.
 (ii) It also defines the word 'quarter' in subsection (3) as the period
     of three months which begins on the first day of January, April,
     July or October in any year.

3 *Regulations regarding felling licences*

(a) Applications for felling licences should be made to the
Forestry Commission and sent to the conservator in whose
conservancy the trees which are to be felled are growing. The
addresses of the various conservancies will be found in Chapter XIX.
(b) The applicant must be the person who owns the land on
which the trees are growing and who has the right to fell them.
(c) Felling licences remain in force for the period stated on the
licence, but this will not be less than one year from the date of issue.

(*d*) Where a licence is refused, the owner of the trees may be entitled to compensation. If, however, the applicant serves a notice on the Commission requiring it to buy the trees, compensation will not be payable.

(*e*) A licence may be issued subject to conditions regarding the replanting of the land on which the trees are growing or such other land as may be agreed, and for the maintenance of the trees which have been replanted.

(*d*) Any person who fells a tree without a felling licence, when a licence is necessary, will be subject to:
  (i) a fine not exceeding £1,000 *or*
  (ii) twice the amount which the court considers to be the value of the tree,
whichever is the larger amount.

4 *Felling directions*

(*a*) Although the Forestry Commission have powers to control the felling of trees, they also have powers to give directions to the owners of trees to fell them, if they consider it expedient to do so.

(*b*) The following are not subject to felling directions:
  (i) Fruit trees, trees growing in orchards, gardens or churchyards or trees growing in public open space.
  (ii) Trees growing on land which is subject to a forestry dedication covenant or agreement.
  (iii) Trees which are managed in accordance with a plan approved by the Commission but which are not covered by a dedication covenant or agreement.

5 *Notices and appeals*

(*a*) Where any work which is required to be carried out under a felling licence or a felling direction, is not done, the Forestry Commission may give notice to those concerned, requiring them to carry out the work.

(*b*) If after serving notice the work is not completed, the Commission may carry out the work. Furthermore a person failing to do the work may be subject to a fine.

(*c*) A person who receives a notice can appeal to the Minister of Agriculture, Fisheries and Food or the Secretaries of State for Wales or Scotland on the grounds that the work has been carried out or that the work required by the notice is not required by the conditions or directions.

Note. The foregoing is a brief summary of the position regarding felling licences and felling directions. When dealing with an actual case, reference should be made to the appropriate legislation.

CHAPTER XVII

Tree Preservation Orders

1 *Introduction*

(*a*) This chapter provides only a general summary of the various matters which concern tree preservation orders, and for more detailed information, reference should be made to the following:

(i) The statutes and statutory instruments which are given in paragraph (*b*) below.

(ii) A leaflet issued by the Department of the Environment and the Welsh Office entitled *Protected Trees – A Guide to Tree Preservation Procedures*. Copies may be obtained from the Department of the Environment, Victoria Road, South Ruislip, HA4 0N2.

(iii) The local planning authority.

(*b*) In 1989, the following statutes and Statutory Instruments which are concerned either wholly or partly, with preservation orders were in force:

(i) *Statutes*

Forestry Act 1947
　Section 15 and as amended
Town and Country Planning Act 1971
　Sections 59–62, 102, 103, 174, 175, 276
Town and Country Amenities Act 1974
　Sections 8 and 10
Local Government, Planning and Land Act 1980
　Schedules 15 and 34
Town and Country Planning Act 1984
　Section 2
Town and Country Planning (Amendment) Act 1985

(ii) *Statutory Instruments*

1969 No. 17　The Town and Country Planning (Tree Preservation Order) Regulations 1969
1975 No. 148　The Town and Country Planning (Tree Preservation Order) (Amendment) and (Trees in

Conservation Areas) (Exempted Cases) Regulations 1975

1981 No. 14 The Town and Country Planning (Tree Preservation Order) (Amendment) Regulations 1981

1988 No. 963 The Town and Country Planning (Tree Preservation Order) (Amendment) Regulations 1988

2 Scope and extent

(a) Under the Town and Country Planning Act 1971, Section 60, a local planning authority can make a tree a preservation order which may apply to individual trees, groups of trees or woodlands as described in the order.

(b) The order may prohibit the felling, uprooting, mutilating, wilfully damaging or destroying the trees to which the order applies unless the consent of the planning authority has been obtained.

(c) The Town and Country Amenities Act 1974, Section 8, deals with trees in a conservation area which are not covered by a preservation order.

3 Making an order

(a) Under Section 60(4) of the Planning Act of 1971, a preservation order will not take effect until it has been confirmed by the Secretary of State unless it is unopposed or any objections to it are withdrawn, when such confirmation is not required.

(b) However, if the local planning authority considers that a preservation order should come into operation immediately, it can act accordingly. In such cases the order will continue for a period of six months or until it has been confirmed, whichever is the sooner.

4 The care of protected trees

(a) After a preservation order has been made, the owner of the tree is still responsible for its condition and for any damage that it might cause.

(b) At the same time, he cannot carry out any work on a protected tree without obtaining permission from the planning authority except in the following cases:

(i) Dead, dying or dangerous trees.

(ii) When required to do so under an Act of Parliament or when requested to do so by a government department or other specified authority.

(iii) Where a tree is obstructing development for which planning permission has already been given.

(iv) Fruit trees.

(v) In order to prevent or overcome a nuisance.

(vi) When the trees are covered by a plan of operations which has been approved by the Forestry Commission and is in force, or if the Commission has already issued a felling licence.

(c) However, in order to safeguard his position, an owner would be prudent to give the planning authority not less than five days' notice of any proposal to fell a protected tree. The only exception to this would be in the case of an emergency.

(d) Before carrying out any work on a tree which is the subject of a preservation order, permission should be obtained from the planning authority.

5 *Penalties*

(a) If the owner of a protected tree, fells or uproots it or does anything to it that can be construed as wilful damage which would probably destroy it, he can be fined up to £2,000 or twice the amenity value of the tree, whichever is the greater.

(b) For other offences, such as the contravention of a tree preservation order, an owner may be fined up to £1,000 and £5 per day for continuing offences.

6 *Replanting*

(a) If the owner of a protected tree fells or destroys it, he will be required to replant it, except in the following cases:

(i) Where the planning authority does not require a replacement.

(ii) In a woodland, where the tree is dangerous, dead or dying.

(b) Replanting must also be carried out:

(i) When permission is granted to fell a protected tree subject to a replanting condition.

(ii) When permission is given in respect of a woodland which is covered by a preservation order. This does not apply, however, where the proposed felling is in the form of a thinning, is in connection with an approved development or where replanting conditions have been withdrawn.

7 *Compensation*

Under the Town and Country Planning Act 1971, an owner can claim compensation from the local planning authority as follows:

(a) *Section 174*

For any damage or loss which occurs if his request for permission to carry out work on a protected tree is refused.

(b) *Section 175*

Where permission has been given to fell a protected woodland, subject to it being replanted, but the Forestry Commmission will not authorize the payment of a grant because the conditions laid down by the planning authority conflict with the growing of timber.

CHAPTER XVIII

Records and Labelling

A *Records*

1 *General*

(*a*) Information relating to specimen trees should be carefully recorded so that their case history is readily available.

(*b*) As regards individual trees, the minimum information recorded will be:

(i) Tree or arboretum number.
(ii) Species.
(iii) Date planted.
(iv) Source from which obtained, i.e. nursery history.
(v) Any damage suffered, with dates.
(vi) Date and details on any tree surgery carried out.
(vii) Periodical measurements.
(viii) Date of death and/or felling.
(ix) Replacement species.

(*c*) In addition to information regarding individual trees, records should also be kept of climatic and other events which may affect the trees concerned. Examples of such events are:

(i) Very heavy frosts.
(ii) Gales or high wind and their effect.
(iii) Heavy falls of snow and resulting damage.
(iv) Flooding.
(v) New areas taken over for planting.
(vi) Areas given up for building, road construction, laying of underground pipes, cables and so on.

(*d*) The maintenance of detailed records can provide valuable information on such matters as the reasons for the success or failure of a tree, its rate of growth, the suitability of the site and any damage that it may have suffered.

(*e*) All trees should be allotted a number, and plans prepared on which the position of individual trees may be shown simply by their number. Where there are many trees growing together in a park, cemetery, arboretum or playing fields, small brass labels about 3 cm (1½ in.) long bearing the tree number can be attached to each tree, so as to avoid confusion or errors. These should be fixed in an inconspicuous position about six to seven feet above ground level, so as to safeguard them as far as possible from vandalism.

2 *Methods of recording information*

(*a*) The information referred to above may be recorded in three ways:
(i) In a bound book.
(ii) In a loose-leaf book.
(iii) By card index.

(*b*) Each of these methods has advantages and disadvantages. A bound book is probably the most secure method, but all entries must be made by hand, while the removal of unwanted or obsolete pages is difficult. Pages in loose-leaf books can be rearranged without difficulty, and entries may be typed on the pages which are temporarily removed, but on the other hand there is some risk of losing a page. A card index is in many ways similar to a loose-leaf book but is far less portable.

(*c*) Records should be written up when any event of importance occurs. Apart from this it is wise to go through all records at least once a year, as for example at the end of the planting season on 31 March or at the beginning of the calendar year on 1 January.

3 *Measurement records*

(*a*) The recording of the measurements of individual trees can be a source of considerable interest, but when these measurements are taken at regular intervals they can also provide valuable information as to girth, height and crown growth.

(*b*) Girth measurements should be taken at 1.3 m (5 ft) above ground level.

(*c*) Heights should be calculated with one of the height measuring instruments which are now available.

(*d*) When a tree is windblown or is felled, care should be taken to ensure that the height is accurately measured and compared with the latest instrumental measurement. The results should be entered in the record book under the tree to which they apply.

B *Labelling*

1 *The purpose of labels*

(*a*) The purpose of a tree label is to give as much information as possible in the most concise form. Correct labelling can add considerably to the interest in the trees concerned and also to their educational value.

(*b*) The information which a label provides can be set out in different ways, but one method which has been found to be satisfactory is given below. In its briefest form the label should provide the following facts:

(i) Arboretum or catalogue number of the tree.
(ii) Common name in English.
(iii) Botanical name in Latin.
(iv) Zone or country of origin.

(*c*) A label so inscribed could appear thus

118 DOUGLAS FIR PSEUDOTSUGA MENZIESII WESTERN NORTH AMERICA

2 *Types of Labels*

(*a*) Tree labels may be made from several materials. In the past, porcelain labels have been used, but at the present time the following are generally adopted:

Cast metal
Inscribed lead
Plastic

(*b*) Cast metal labels, although durable, are not always readily available for the rarer trees.

(*c*) Inscribed or stamped lead labels have been used frequently in the past, but unfortunately they are often damaged by grey squirrels.

(*d*) Plastic labels are most commonly used at the present time. They can be obtained in various background colours and can be inscribed without difficulty by means of an engraving machine.

(*e*) The attachment of a label to a tree is not always an easy matter and the following points should be borne in mind:

(i) Unless they are firmly attached to the tree they may be moved or lost.

(ii) A healthy tree increases in girth each year, and if a label is attached by nailing or by wire, in course of time the nail, wire or the label itself may become engulfed by the expanding trunk.

(iii) A brass screw which can be unscrewed one turn each spring or autumn is probably the best method of fixing.

(iv) Labels should be placed at least six feet above ground level, so that they are out of reach of those who enjoy removing them or breaking them.

CHAPTER XIX

The Forestry Commission

1 The formation of the Forestry Commission

(*a*) Although the management of the Forest and Land Revenues was transferred to the Commissioners of Woods and Forests in 1810, their duties were largely concerned with the ancient royal forests, of which the New Forest and the Forest of Dean were the most important. It was not until 1924 that the forest areas held by the Commissioners of Woods were transferred to the Forestry Commissioners under the Forestry (Transfer of Woods) Act 1923.

(*b*) After the outbreak of war in 1914, the need for adequate supplies of home-grown timber was soon realized. As submarine warfare gained momentum and the losses of allied shipping mounted, the country became increasingly dependent on home-grown timber. As a result large areas were felled, but since the national forests were so limited in extent the heaviest demands were made on privately owned woodlands.

(*c*) In 1916 a Forestry Sub-Committee of the Reconstruction Committee was appointed under the chairmanship of the Rt Hon. F. D. Acland, and 10 months later its report was published. This is generally referred to as the Acland Report, and it was in the light of the Committee's recommendations that the Forestry Commission was established after the passing of the Forestry Act 1919.

2 The development of the Commission

(*a*) Under the chairmanship of Lord Lovat the Commission made steady progress, and in 1927 another Forestry Act was passed which increased the number of Commissioners from eight to ten.

(*b*) During the 1939–45 war history repeated itself and further heavy fellings were necessary, the majority of which, once again, occurred in private woodlands. In 1943 the Forestry Commissioners prepared and presented a report to the Chancellor of the Exchequer entitled *Post-War Forest Policy*, and this was followed by a supplementary report, *Post-War Forest Policy, Private Woodlands*, in 1944. These reports laid the foundations of the forest policy for this country during the first 20 years after the war.

(*c*) The Forestry Act 1945 reconstituted the Commission, enacted certain provisions as to the acquisition of land, and made certain amendments to the Forestry Acts 1919 to 1927.

(*d*) In 1947 a further Forestry Act was passed which covered certain aspects of Forestry dedication covenants and the restriction of compulsory powers regarding the acquisition of dedicated land.

(*e*) Four years later the Forestry Act 1951 set out the law regarding the felling of trees and the licences which were necessary, with a view to safeguarding adequate stocks of standing timber. It also reconstituted Regional Advisory Committees and the Home-Grown Timber Advisory Committee as statutory committees.

(*f*) In 1965 the Forestry Commission was reorganized to a considerable extent, and this was followed by the Forestry Act 1967, which consolidated the provisions of the earlier Forestry Acts and repealed those of 1919, 1923, 1927, 1945, 1947 and 1951.

(*g*) On 24 October 1973 new proposals were announced relating to the payment of grants, a new Dedication Scheme and the membership of the Regional Advisory Committees, while a statement made on 5 July 1974 gave further details of these proposals. Subsequently the Commission redefined its objectives, which were set out in the Fifty-Fourth Annual Report.

(*h*) The Forestry Act 1979 restated the powers of the Commissioners to make grants and loans, amended parts of S.9 and S.43 of the Forestry Act 1967 by substituting metric measure for Imperial and repealed part of S.9(6).

(*i*) The Forestry Act 1981 dealt with the disposal of the Commissioners' property, certain financial matters and an increase in the number of Forestry Commissioners from 9 to 10 excluding the Chairman. In October 1981, the Basis III Dedication Scheme and the Small Woods Scheme were replaced by the Forestry Grant Scheme.

(*j*) A new Government policy for broadleaved woodlands was announced in July 1985 and the Broadleaved Woodland Grant Scheme was introduced in October of that year. It also became the duty of the Commission, under the Wildlife and Countryside (Amendment) Act 1985 (which amended S.1 of the Forestry Act 1967), to try to reach a satisfactory balance between the interests of forestry and the environment.

(*k*) In April 1988 the Woodland Grant Scheme was introduced which replaced the two schemes referred to in the previous paragraphs.

3 *The organization of the Commission*

The following is a brief account of the organization of the Forestry Commission as on 1 April 1988.

(*a*) *Ministerial responsibility*
Responsibility for forest policy is shared by the Minister of Agriculture, Fisheries and Food, the Secretary of State for Scotland and the Secretary of State for Wales.

(*b*) *The Commissioners*
 (i) Under the Forestry Act 1967, as amended by the Forestry Act 1981, there are eleven Commissioners, of whom seven, including the Chairman, are only part-time. The remaining four, who are full-time, are the Director-General and three senior officers of the Commission.
   (1) At least three should have special knowledge and experience of forestry.
   (2) At least one should have scientific attainments and a technical knowledge of forestry.
   (3) At least one should have a special knowledge and experience of the timber trade.
 (ii) The names of the Commissioners, senior members of the staff, conservators and members of the committee and other bodies appointed by the Commissioners are published in the Annual Report and Accounts of the Forestry Commission.

(*c*) *Headquarters organization*
Chairman
Deputy Chairman and Director-General
Three Executive Commissioners responsible respectively for:
  Administration and Finance
  Operations
  Private Forestry and Development
Secretary
Director, Harvesting and Marketing
Director, Research
Director, Planning and Surveys
Director, Estate Management
Director, Private Forestry and Services

Director, Personnel
  Controller of Finance
  Head of Silviculture
  Head of Data Processing
  Chief Engineer
  Head of Information

Note. Early in 1975 the headquarters of the Commission was transferred from London and Basingstoke to Edinburgh. The new Headquarters Office at 231, Corstorphine Road, Edinburgh EH2 7AT (telephone 031 334 0303) was formally opened by the Secretary of State for Scotland on 16 May 1975.

(*d*) *Forest Research Stations*
The Commission has two research stations:
(i) Forestry Commission Research Station, Alice Holt Lodge, Wrecclesham, Farnham, Surrey, GU10 4LH (telephone 0420 22255).
(ii) Forestry Commission Northern Research Station, Roslin, Midlothian, EH25 9SY (telephone 031 445 2176).

(*e*) *Conservancies*
For purposes of administration, England, Scotland and Wales are divided into regions known as conservancies, and until April 1985 there were a total of eleven, of which five were in England, four in Scotland and two in Wales. Prior to 1969, the New Forest and Dean Forest formed two separate charges under their own Deputy Surveyors, but in that year they became part of the South East and South West Conservancies respectively. However, from 1 April 1985 these eleven conservancies were reduced to seven, while in April 1988, further minor alterations were made to the Mid Scotland and South Scotland Conservancies in respect of the Motherwell District. Every conservancy is sub-divided into Forest Districts which vary in number from seven in the South Scotland Conservancy to fourteen in the Wales Conservancy, each having its own District Office. The extent of the seven conservancies with the location of the Conservancy Offices, as in April 1988, are given below.

(1) *North England*
  Extent: The counties of Northumberland, Tyne and Wear, Durham, Cleveland, Cumbria, Lancashire, Cheshire, Merseyside, Greater Manchester, West Yorkshire, North Yorkshire, South Yorkshire and Humberside except the districts of Glanford, Scunthorpe, Grimsby and Cleethorpes.
  Office: 1A Grosvenor Terrace, York YO3 7BD.

(2) *East England*
  Extent: The counties of Lincolnshire, Bedfordshire, Norfolk, Suffolk, Essex, Hertfordshire, Buckinghamshire, Northamptonshire, Cambridgeshire, Oxfordshire and Leicestershire. That part of the Staffordshire Moorlands District of Staffordshire which falls within the Peak District National Park. That part of Humberside comprising the districts of Glanford, Scunthorpe, Grimsby and Cleethorpes. Greater London, Nottinghamshire, Derbyshire, Berkshire, Hampshire, Kent, Surrey, East Sussex, West Sussex and Isle of Wight.
  Office: Great Eastern House, Tenison Road, Cambridge CB1 2DU.

(3) *West England*
  Extent: The counties of Cornwall, Devon, Somerset, Dorset, Wiltshire, Avon, Gloucestershire, Hereford and Worcestershire, Warwickshire, West Midlands, Shropshire and the whole of Staffordshire except that part of the Staffordshire Moorlands District which falls within the Peak District National Park.
  Office: Avon Fields House, Somerdale, Keynsham, Bristol BS18 2BD.

(4) *Wales*
  Office: Victoria House, Victoria Terrace, Aberystwyth, Dyfed SY23 2DQ.

(5) *North Scotland*
  Extent: The whole of the Western Isles Authority area, Shetland Isles Authority Area and Orkney Isles Authority area. The whole of the Highland Region area with the exception of that portion of Lochaber District east of Loch Linnhe and south of Loch Leven, the River Leven, the Blackwater Reservoir and the Blackwater, to where the Highland Region boundary reaches the river of Lochan A'chlaidheimh. Strathclyde Region: the Isles of Mull, Iona, Coll and Tiree only. The whole of the Grampian Region.
  Office: 21 Church Street, Inverness IV1 1EL.

(6) *Mid Scotland*
  Extent: That portion of the Lochaber District of Highland Region east of Loch Linnhe and south of Loch Leven, the River Leven, the Blackwater Reservoir and the Blackwater, to where the Highland Region boundary reaches the river at

Lochan A'chlaidheimh. Central Region, Tayside Region and Fife Region. That part of Strathclyde Region comprising the districts of Argyll and Bute – except the Isles of Mull, Iona, Coll and Tiree – Dumbarton, Clydebank, Bearsden and Milngavie, Strathkelvin, Cumbernauld, Monklands, part of Motherwell (see South Scotland), Hamilton, City of Glasgow, Inverclyde, Renfrew, that portion of East Kilbride District north of a line from Routen Burn to Eaglesham, and that portion of Eastwood District north and west of the B764 from Eaglesham to the district boundary.

Office: Portcullis House, 21 India Street, Glasgow G2 4PL.

(7) *South Scotland*
  Extent: The regions of Lothian, Borders and Dumfries and Galloway. That part of Strathclyde Region comprising the districts of Lanark, Cumnock and Doon Valley, Kyle and Carrick, Cunninghame, Kilmarnock and Loudoun, the portion of East Kilbride District south of a line from Routen Burn to Eaglesham, that portion of Eastwood District south and east of the B764 from Eaglesham to the district boundary, and part of Motherwell District including Shotts, Hartwood, Allanton and the area east of the 'C' road between Allanton and Carluke.

  Office: Greystone Park, 55/57 Moffat Road, Dumfries DG1 1NP.

4 *The committees of the Commission*

The following are the principal committees appointed by or in conjunction with the Commission.

(*a*) *National Committees*
 (i) Three National Committees, for England, Scotland and Wales respectively, were appointed under the provisions of the Forestry Act 1945, when they replaced the earlier Consultative Committees. Their functions are now purely advisory.
(ii) In 1987 the Committee for England consisted of seven members, of whom four were Commissioners and the remainder chairmen of the three English Regional Advisory Committees. The Committee for Scotland was composed of three Commissioners and three chairmen of the Advisory Committees, while the Committee for Wales comprised two Commissioners and four independent members.

(b) *Regional Advisory Committees*
  (i) These were formed in 1946, one Committee being set up in each conservancy, the object being to provide a link between the Conservator and those in the conservancy who are interested in forestry.
  (ii) In October 1974 the membership of the Committees was enlarged so as to include agricultural, amenity and local planning interests as well as those of woodland owners and timber merchants.

(c) *Home-Grown Timber Advisory Committee*
  (i) This Committee was originally formed in 1939 and in 1987 it numbered twenty-five members, seven of whom were appointed independently. The remaining eighteen were representatives of woodland owners, timber merchants, British Coal, the particleboard and pulpwood industries and timber research and development.
  (ii) The function of the Committee is to advise the Commissioners on all matters relating to home-grown timber.
  (iii) The Committee can also appoint sub-committees to report on any particular aspect of home timber production, and in 1987 two had been appointed – the Supply and Demand Sub Committee and the Technical Sub-Committee.

(d) *Advisory Committee on Forest Research*
The Committee, which was appointed in 1929, is an internal committee of the Forestry Commission and advises the Director of Research on the quality and direction of research carried out by the Commission.

(e) *Forestry Research Coordination Committee*
This Committee was established in 1982, and its terms of reference may be summarized as follows:
  (i) To identify and define forestry research needs.
  (ii) To advise on research requirements and priorities.
  (iii) To stimulate forest research, the exchange of information and collaboration between research organizations and individuals.
  (iv) To publish the results of such research.
  (v) To encourage the financing of research proposals.

(f) *Forestry Training Council*
  (i) Although this body is not a committee, it has been included, since it is appointed by the Commission.
  (ii) It was set up in 1971 after forestry interests had withdrawn from the Agricultural, Horticultural and Forestry Industry

Training Board, and is now responsible for providing training courses for personnel both in the Commission and in the private sector.

(g) *Forestry Safety Council*
(i) The Council occupies a similar position to that of the Forestry Training Council in that it is not a committee but at the same time is appointed by the Forestry Commission.
(ii) It was established in 1974 for the purpose of promoting safety in forestry.

5 *The objectives of the Commission*

The Forestry Commission has two distinct parts to play. First, as the Forestry Enterprise, it is responsible for the management of the forests and land under its control, and second, it is the Forest Authority. As such it advises on forest policy, carries out research, controls felling, deals with grants and so on. Its objectives, which are set out in *Forest Facts 2*, issued by the Commission, are as follows:

(a) *As Forestry Enterprise*
(i) To develop its forests for the production of wood for industry, by extending and improving the forest estate.
(ii) To manage its estate economically and efficiently and to account for its activities to Ministers and Parliament.
(iii) To protect and enhance the environment.
(iv) To provide recreational facilities.
(v) To stimulate and support employment and the local economy in rural areas by the development of forests, including the establishment of new plantations, and of the wood-using industry.
(vi) To foster a harmonious relationship between forestry and other land use interests, including agriculture.

(b) *As Forestry Authority*
(i) To advance knowledge and understanding of forestry and trees in the countryside.
(ii) To develop and ensure the best use of the country's forest resources and to promote the development of the wood-using industry and its efficiency.
(iii) To endeavour to achieve a reasonable balance between the interests of forestry and those of the environment.

(iv) To undertake research relevant to the needs of forestry.
(v) To combat forest and tree pests and diseases.
(vi) To advise and assist with safety and training in forestry.
(vii) To encourage good forestry practice in private woodlands through advice and schemes of financial assistance and by controls on felling.

6 *The publications of the Commission*

One of the first steps to be taken by the Commission after its formation in 1919 was to publish information on forestry matters, an early example being a leaflet on pine weevils, which appeared in February 1920. Since then, the many publications issued in the Commission's various series have provided an outstanding source of reliable information. A further account of these will be found in Chapter XXIII.

7 *The Arboricultural Advisory and Information Service*

A short account of the Service and its work will be found in Chapter XXI.

CHAPTER XX

Arboricultural Education, Training and Research

1 *Introduction*

(*a*) Although arboriculture was practised long before forestry, it did not make the same technical progress, and as a result education did not achieve the same standard as it did in forestry.

(*b*) The first effective step to recognize arboriculture as a separate subject from forestry was taken by the Royal Forestry Society of England, Wales and Northern Ireland in 1958, when the Society formed its own arboricultural section. This has now been superseded, and arboricultural meetings are held by the various Divisions each year. In 1966 the Royal Scottish Forestry Society formed an arboricultural group, but this has since been discontinued.

(*c*) In 1964 two associations were formed, the Association of British Tree Surgeons and Arborists, and the Arboricultural Association. These two bodies joined forces in 1974 under the title of The Arboricultural Association, and further information relating to it will be found in Chapter XXI.

(*d*) The first examinations in arboriculture were held by the Royal Forestry Society of England, Wales and Northern Ireland in 1958, and since then the Society has conducted arboricultural examinations each year.

2 *Examinations*

(*a*) In 1989 there were three bodies that conducted examinations in arboriculture, and details of these are given below.

(*b*) The Royal Forestry Society of England, Wales and Northern Ireland holds two examinations:

(i) *Certificate in Arboriculture*
This is a craft level qualification and is the first step for those who wish to proceed to the Society's advanced Professional Diploma in Arboriculture.

(ii) *Professional Diploma in Arboriculture*

    (1) This was originally known as Master of Arboriculture but from 1990 the title has been changed to Professional Diploma in Arboriculture. However, those who had already passed this examination can continue to use the designation of Master of Arboriculture.

    (2) The examination provides an advanced qualification which can lead to managerial appointments.

(iii) The syllabus for both of these examinations can be obtained from The Royal Forestry Society of England, Wales and Northern Ireland, 102 High Street, Tring, Hertfordshire HP23 4AH.

    (c) The Arboricultural Association provides a Technician's Certificate in Arboriculture at a level between the two examinations of the Royal Forestry Society. Details of this Certificate are available from The Arboricultural Association, Ampfield House, Ampfield, Romsey, Hampshire SO51 9PA.

    (d) The City and Guilds of London Institute offers a Certificate in Tree Surgery, at both Craft and Foreman level, at Merrist Wood College of Agriculture. Full details can be obtained either from the Institute, 76 Portland Place, London W1N 4AA, or from Merrist Wood College of Agriculture, Worplesdon, Guildford, Surrey GU3 3PE.

3 *Training*

    (a) Training in arboriculture is now available at the Royal Botanic Garden, Edinburgh, and at a number of agricultural colleges. Some of these hold courses which are solely concerned with arboriculture, while others include it as part of a broader curriculum such as amenity horticulture.

    (b) The following is a list of colleges or centres at which courses in arboriculture or courses which include a certain amount of arboriculture, are available. Full details of these can be obtained from the addresses given below:

Askham Bryan College of Agriculture and Horticulture,
    Askham Bryan, York YO2 3PR

Cannington College of Agriculture and Horticulture,
    Cannington, Bridgwater, Somerset TA5 2LS

Capel Manor Horticultural and Environmental Centre,
    Bullsmore Lane, Waltham Cross, Herts. EN7 5HR

Cheshire College of Agriculture,
    Reaseheath, Nantwich, Cheshire CW5 6DF

Cumbria College of Agriculture and Forestry,
  Newton Rigg, Penrith, Cumbria CA11 0AH
Durham College of Agriculture and Horticulture,
  Houghall, Durham, DH1 3SG
Holme Lacey College,
  Holme Lacey, Hereford HR2 6LL
Knowsley Central Tertiary College,
  Rupert Road, Roby, Merseyside L36 9TD
Lancashire College of Agriculture and Horticulture,
  Myerscough Hall, Bilsborrow, Preston PR3 0RY
Lincolnshire College of Agriculture and Horticulture,
  Caythorpe Court, Caythorpe, Grantham, Lincs. NG32 3EP
Merrist Wood College of Agriculture and Horticulture,
  Worplesdon, Guildford, Surrey GU3 3PE
Royal Botanic Garden,
  Inverleith Row, Edinburgh EH3 5LR.

4 *Research*

Research into various matters concerned with, or affecting,
arboriculture is carried out by the following Government institu-
tions, while some research projects are also undertaken by a
number of universities:

(*a*) *The Forestry Commission*
  (i) There are two research stations:
        Forestry Commission Research Station, Alice Holt Lodge,
          Wrecclesham, Farnham, Surrey GU10 4LH
        Forestry Commission Northern Research Station, Roslin,
          Midlothian EH25 9SY.
 (ii) In February 1988 the *Report of the Review Group on Arboricul-
      ture* was published under the direction of the Forestry Research
      Coordination Committee. This covers arboriculture as a whole,
      considers its position in relation to forestry, and puts forward
      recommendations for the future. It should be read by all who
      are concerned with the subject.
(iii) Further information on the research committees of the
      Forestry Commission will be found in Chapter XIX.

(*b*) *East Malling Research Station*
  (i) This is situated at East Malling, Maidstone, Kent ME19 6BJ.
 (ii) It is part of the Institute of Horticultural Research, and in
      addition to its work on fruit trees, it undertakes research
      relating to ornamental trees, shrubs and nursery stock.

(*c*) *Institute of Terrestrial Ecology*
   (i) This is one of the component institutes of the Natural Environment Research Council (NERC).
  (ii) It comprises two divisions:

   Institute of Terrestrial Ecology (North), Bush Estate, Penicuik, Midlothian EH26 OQB.

   Institute of Terrestrial Ecology (South), Monks Wood Experimental Station, Abbots Ripton, Huntingdon PE17 2LS.

 (iii) The wide-ranging research undertaken by the Institute includes investigations into tree health, planting on open-cast sites, and other matters relating to the wise use of natural resources.

(*d*) *Long Ashton Research Station*
   (i) The Long Ashton Research Station, Long Ashton, Bristol BS18 9AF operates as part of the Institute of Arable Crops Research.
  (ii) For many years research has been carried out on the cultivation of uses of willows for basket making, windbreaks, amenity and, more recently, the production of willow biomass.

CHAPTER XXI

Organizations Concerned with Arboriculture

1 *The Arboricultural Advisory and Information Service*

(*a*) The Service was established in 1976 by the Forestry Commission under contract to the Department of the Environment. It is operated by the Commission under the direction of the Arboricultural Advisory and Information Officer.

(*b*) The purpose of the Service is 'to disseminate research results, monitor research requirements and communicate to the arboricultural and allied industries, sound cultural practices'.

(*c*) The Service prepares and issues the following:

(i) *Arboricultural Research Notes*
These provide the results of current research and information on the cultivation and management of trees.

(ii) *Assistance with Arboricultural Reading*
This is a valuable guide to both technical and practical articles which have been published on arboricultural and related matters. Photocopies of articles which are not readily available can also be supplied.

(iii) *Arboricultural Leaflets*
After a research project has produced results that can be finally accepted, the recommendations are published as an Arboricultural Leaflet. Further information on these will be found in Chapter XXIII.

(*d*) The Service also provides advice, and when this is given by correspondence or telephone, no charge is made. Site visits can be arranged, but a fee is payable for these and details are available from the Arboricultural Advisory and Information Officer.

(*e*) Correspondence should be addressed to The Arboricultural Advisory and Information Service, Forest Research Station, Alice Holt Lodge, Wrecclesham, Farnham, Surrey GU10 4LH.

2 *Associations and societies*

This section contains brief particulars of a number of associations, societies and other bodies which, to a greater or lesser extent, are concerned with, or have an interest in, arboriculture. They are arranged in alphabetical order, and further information regarding them can be obtained from the addresses which are given at the end of each entry.

(*a*) *Arboricultural Association*
  (i) The Association was founded in 1964 with the object of raising the standards of arboriculture and to assist in the education of those who choose arboriculture as a career. In 1974 it amalgamated with the Association of British Tree Surgeons and Arborists but the title of 'Arboricultural Association' was retained. It is now recognized as the leading body, which represents professional arboriculture in the United Kingdom.
  (ii) The Association publishes a journal, the *Arboricultural Journal*, and an information sheet entitled *News* every quarter.
  (iii) In 1988 the Association formed the Arboricultural Safety Council, and further information on this will be found in Chapter VIII.
  (iv) The membership of the Association comprises Fellows, Associates, Affiliates, Ordinary Members, Corporate Members and Students.
  (v) Further details of the Association can be obtained from the Secretary, Arboricultural Association, Ampfield House, Ampfield, Romsey, Hampshire SO51 9PA.

(*b*) *International Dendrology Society*
  (i) The objects of the Society, which was formed in 1952, are to encourage and advance the study and cultivation of trees and shrubs and to safeguard those species which are rare or endangered.
  (ii) The Society publishes a *Year Book* and *Newsletters*.
  (iii) Further information can be obtained from the Secretary, International Dendrology Society, School House, Stannington, Morpeth, Northumberland NE61 6HF.

(*c*) *Landscape Institute*
  (i) This is a professional institute for landscape architects, scientists and managers.
  (ii) It was previously known as the Institute of Landscape Architects.

(iii) The Institute's address is 12 Carlton House Terrace, London SW1Y 5AH.

(*d*) *Men of the Trees*
 (i) This body was founded in 1922 and is concerned with amenity, specimen trees and woodlands.
 (ii) It publishes a journal entitled *Trees* and also a newsletter, both of which are issued twice yearly.
(iii) Those wishing to join may do so either as members or life members.
(iv) Further details are available from the Secretary, Crawley Down, Crawley, Sussex RH10 4HL.

(*e*) *The Royal Forestry Society of England, Wales and Northern Ireland*
 (i) Founded in 1882 as the English Arboricultural Society, the Society holds arboricultural meetings as well as those devoted to forestry.
 (ii) It publishes the *Quarterly Journal of Forestry* as well as various studies on different aspects of forestry.
(iii) There are several classes of membership, ranging from woodland owners to students.
(iv) Full particulars can be obtained from the Society's headquarters at 102 High Street, Tring, Hertfordshire HP23 4AH.

(*f*) *The Royal Horticultural Society*
 (i) The Society, which was founded in 1804, is primarily concerned with the science and practice of gardening, but it is also interested in certain aspects of arboriculture. The Society's garden at Wisley contains a large collection of trees and shrubs.
 (ii) The journal of the Society, which is entitled *The Garden*, is published every month.
(iii) Information regarding membership is available from the Secretary, 80 Vincent Square, London SW1P 2PE.

(*g*) *Scottish Arboricultural Society*
 (i) The Society was formed in 1987 by a Scottish branch of the Arboricultural Association.
 (ii) The object of the Society is to establish a meeting point for those who are interested in trees and wish to extend their knowledge.
(iii) Membership is open to all who work in, or are interested in, arboriculture and other allied fields, and comprises the following grades: Corporate Members, Professional Members, Non-professional Members and Students.

(iv) More detailed information can be obtained from the Secretary, 4 Knightsbridge Road, Dechmont, West Lothian EH52 6LT.

(*h*) *The Tree Council*
  (i) The Council was formed in 1974 with the following objectives:
    (1) To improve the environment by promoting the planting of trees and their maintenance.
    (2) To extend a knowledge of trees.
    (3) To provide a forum for organizations that are concerned with trees.
    (4) To identify national problems relating to trees and to encourage co-operation.
  (ii) In 1988, the Council was composed of representatives of twenty bodies, who were known as Council Members, and representatives of eleven other bodies known as Consultative Members.
  (iii) Additional information regarding the work of the Council can be acquired from the Secretary, 35 Belgrave Square, London SW1X 8QN.

Note. Grants are available from several bodies in respect of amenity planting, individual trees, small groups of trees and small woods. Details of these will be found in the Appendix.

CHAPTER XXII

Botanic Gardens, Arboreta and Pineta

1 General

(a) Botanic gardens
  (i) This chapter contains a selection of botanic gardens, arboreta and pineta, and it may be helpful to differentiate between a botanic garden and the collections of trees which form an arboretum or pinetum.
  (ii) A botanic garden is primarily concerned with education, and for this reason plants are arranged under their botanical classification rather than for display and ornament. Some botanic gardens undertake plant breeding while most include a library and herbarium.
  (iii) All specimens should be clearly labelled, and additional information concerning them should be available in a card index or other suitable record system.
  (iv) Botanic gardens usually contain a substantial collection of specimen trees.

(b) Arboreta and pineta
  (i) An arboretum is a collection of specimen trees comprising either broadleaved species or a mixture of broadleaves and conifers.
  (ii) A pinetum consists entirely, or predominantly, of conifers.
  (iii) Many of the collections in this country were originally formed by private individuals, as for example Westonbirt Arboretum, and consequently more attention was paid to the appearance and amenity of the collection than to any botanical arrangement.
  (iv) Collections of interesting trees and shrubs are also to be found in some public parks, zoological gardens and other recreational areas.

2 Botanic gardens

The following are among the leading botanic gardens in the United Kingdom and are arranged in alphabetical order:

(a) *Bath*
(i) The Botanical Gardens, which are some 2.8 hectares (7 acres) in extent, were laid out in 1887 and are situated in the Royal Victoria Park. This covers almost 20 hectares (50 acres) and also contains a large number of interesting trees.
(ii) A guide to the Gardens can be obtained from the Bath City Council, Department of Leisure and Tourist Services, The Pump Room, Stall Street, Bath BA1 1LZ.

(b) *Bedgebury*
(i) Although not specifically designated as a botanic garden, Bedgebury was established jointly by the Royal Botanic Gardens, Kew, and the Forestry Commission as the National Pinetum. The object was primarily to provide an alternative site to Kew, where conifers are adversely affected by smoke and other unsuitable conditions.
(ii) The Pinetum covers some 32 hectares (80 acres) and is situated between Cranbrook, Goudhurst and Hawkhurst in West Kent.
(iii) Considerable damage was caused by the great storm of October 1987, when between 25 and 30 per cent of the trees were lost, but action has since been taken to replace these.
(iv) Further information will be found in Forestry Commission publication *A Guide to Bedgebury National Pinetum*.

(c) *Cambridge*
(i) The University Botanic Garden was formed in 1762 but was transferred to its present site, one mile south of the city centre in Bateman Street (off Trumpington Road) in 1846.
(ii) The Garden covers an area of about 16 hectares (40 acres) with entrances in Trumpington Road, Bateman Street and Hills Road, and is open throughout the year.
(iii) It contains systematic collections of notable trees and shrubs.
(iv) A guide may be obtained from the Univesity Botanic Garden, Cory Lodge, Bateman Street, Cambridge CB2 1JF.

(d) *Edinburgh*
(i) The Royal Botanic Garden, Scotland's national botanic garden, was founded in 1670 and moved to its present site in 1823.

(ii) It now covers some 29 hectares (72 acres) and contains over 10,000 species of plants, including a great diversity of trees and shrubs.

(iii) There are two public entrances, one in Arboretum Place, where there is ample parking, and the other beside 7 Inverleith Row, which is served by public transport. The entrance to the Research Department and Offices is at 20 Inverleith Row, Edinburgh EH3 5LR.

(e) Kew

(i) The Royal Botanic Gardens at Kew have been formed over a long period which began in 1759. However, it was after Sir William Hooker had been appointed the first Director in 1841 that the area of the Gardens was increased to over 101 hectares (250 acres).

(ii) Today the Gardens extend to some 130 hectares (320 acres) and include museums, herbarium, library, laboratory and the finest collection of specimen trees and shrubs in the world.

(iii) In 1965, Wakehurst Place, near Ardingly, was leased from the National Trust as an addition to the Royal Botanic Gardens. It includes a large collection of rare trees and shrubs.

(f) Oxford

(i) The Oxford Botanic Garden was founded as a Physic Garden by the Earl of Danby in 1621 and is the oldest botanic garden in Britain.

(ii) It is situated on the south side of the High Street opposite Magdalen College and covers an area of about 2.8 hectares (7 acres).

(iii) A guide to the Garden is available from the Superintendant, Oxford Botanic Garden, High Street, Oxford OX1 4AX.

(iv) In addition to the Botanic Garden, there is an extensive collection of trees and shrubs in the University Parks, the site of which was purchased from Merton College in 1853. A handbook entitled *Guide to the Trees and Shrubs in the University Parks, Oxford* has been published by the Oxford University Press.

(v) A more extensive account of the Botanic Garden, the Parks and College gardens will be found in *The Oxford Gardens* by R. T. Gunther (1912).

Note. Detailed descriptions of 22 botanic gardens and six other plant collections are given in *Collins Guide to the Botanical Gardens of Britain* by Michael Young (1987).

3 *Arboreta and pineta*

The following is a list of some of the outstanding collections of trees which are to be found in the United Kingdom, but owing to lack of space, it has been necessary to limit the number which have been included:

(*a*) *England*
Bedfordshire
    Woburn Abbey, nr. Leighton Buzzard
Berkshire
    Windsor Great Park and Savill Gardens
Cambridgeshire
    Anglesey Abbey, Lode, nr. Cambridge
Cornwall
    Caerhays Castle, Gorran, St Austell
    Glendurgan, Helford, nr. Falmouth
    Trelissick, Feock, nr. Truro
    Tresco Abbey, Isles of Scilly
    Trewithen, Probus, nr. Truro
Cumbria
    Holker Hall, Cark-in-Cartmell, Grange-over-Sands, Lancashire
Derbyshire
    Elvaston Castle, nr. Derby
Devon
    Bicton Pinetum, nr. Budleigh Salterton
    Killerton House, Broadclyst, nr. Exeter
    Knightshayes Court, Bolham, Tiverton
    Reed Hall, University of Exeter, Exeter
Dorset
    Abbotsbury Sub-tropical Gardens, nr. Weymouth
    Forde Abbey, nr. Chard
Gloucestershire
    Batsford Park Arboretum, Moreton-in-Marsh
    Hidcote Manor, nr. Chipping Campden
    Speech House, nr. Coleford, Forest of Dean
    Stanway House, Winchcombe
    Westonbirt Arboretum, nr. Tetbury
Hampshire
    Exbury Gardens, nr. Beaulieu
    Hillier Arboretum, Ampfield, Romsey

New Forest:
   Bolderwood Arboretum, nr. Lyndhurst
   Rhinefield Ornamental Drive, nr. Brockenhurst
Herefordshire
   Eastnor Castle, nr. Ledbury
   Hergest Croft Gardens, Kington, nr. Presteigne
Hertfordshire
   Bayfordbury, nr. Hertford
Lincolnshire
   Brocklesby Park, nr. Immingham
Northumberland
   Cragside, Rothbury, nr. Morpeth
   Howick Gardens, Howick, nr. Alnwick
Surrey
   Winkworth Arboretum, nr. Godalming
   Wisley (Royal Horticultural Society's Garden) between Woking
      and Cobham
Wiltshire
   Bowood Gardens, Calne
   Stourhead, nr. Mere

(b) *Wales*
   Clwyd
      Vivod Forest Garden, nr. Llangollen (Denbighshire)
   Glamorgan
      Duffryn Gardens, St Nicholas, nr. Cardiff
      Roath Park, Cardiff
   Gwynedd
      Bodnant Gardens, Tal-y-Cafn, nr. Colwyn Bay (Denbighshire)
   Powys
      Leighton Redwood Grove, nr. Welshpool (Montgomeryshire)
      Gliffaes Hotel, nr. Crickhowell (Breconshire)
      Stanage Park, nr. Knighton (Radnorshire)

(c) *Scotland*
   Argyll (Strathclyde)
      Benmore Gardens, Dunoon, Firth of Clyde
      Crarae Gardens, Furnace, nr. Inverary
      Inverary Castle, Inverary
      Kilmun Forest Garden, Benmore, Firth of Clyde
   East Lothian
      Smeaton House, East Linton, nr. Dunbar
   Peebles (Borders)
      Dawych Arboretum, Stobo, nr. Peebles

Perthshire (Tayside)
    Blair Castle, Blair Athol, nr. Pitlochry
    Scone Palace, nr. Perth
Ross and Cromarty (Highland)
    Inverewe Garden, Poolewe, nr. Ullapool

(d) *Northern Ireland*
  County Down
    Castlewellan Park, nr. Newcastle
    Rowallane Gardens, Saintfield, nr. Belfast

CHAPTER XXIII

Books, Manuals and Periodicals

1 Books

(a) The following books, the majority of which have been published since 1980, deal with various aspects of arboriculture:

(i) *Diseases*

Phillips, D. H. and Burdekin, D. A., *Diseases of Forest and Ornamental Trees* (Macmillan, 1982).

(ii) *Identification*

Bean, W. J., *Trees and Shrubs Hardy in the British Isles*, 8th edn, 4 vols (John Murray, 1970–80).

Clarke, D. L., *Supplement* to 8th edn of W. J. Bean (above) (John Murray, 1988).

Cooper, M. R. and Johnson, A. W., *Poisonous Plants and Fungi* (HMSO, 1988).

Mitchell, A. F., *A Field Guide to the Trees of Britain and Northern Europe* (Collins, 1974).

Mitchell, A. F. and Jobling, J., *Decorative Trees for Country, Town and Garden* (HMSO, 1984).

Rushforth, K. D., *Conifers* (Christopher Helm, 1987).

(iii) *Planting*

Gruffydd, J. St Bodfan., *Tree Form, Size and Colour* (E & F. N. Spon, 1987).

Rushforth, K. D., *The Hillier Book of Tree Planting & Management* (David & Charles, 1987).

(iv) *Tree surgery*

Bridgeman, P. H., *Tree Surgery* (David & Charles, 1976).

Shigo, A. L., Vollbrecht, K. and Hvass, N., *Tree Biology and Tree Care* (Honey Brothers Ltd, New Pond Road, Peasmarsh, Guildford, Surrey GU3 1JR).

(b) The two books referred to below are unfortunately out of print but they may be found in a library or second-hand bookshop:

Caborn, J. M., *Shelterbelts and Windbreaks* (Faber & Faber, 1965).

Colvin, B. and Tyrwhitt, J. (with line drawings by S. R. Badmin), *Trees for Town and Country* (Lund Humphries, 1947).

2 Manuals

In this context, the term 'manual' is intended to cover those publications which normally have fewer pages than a book and do not have hard covers.

(*a*) *Forestry Commission publications*
  (i) In November 1986, it was decided to reorganize the various categories of publications which were issued by the Commission, and that in future the majority of the priced technical publications would appear in one of the following forms: Bulletins, Handbooks or Field books.
 (ii) However, in addition to these, the Commission is continuing to issue Arboricultural Leaflets and Occasional Papers.
(iii) Many of the Commission's publications cover subjects which are of interest to arboriculturalists as well as to foresters, such as those which deal with beech bark disease and air pollution. Space does not allow further reference to these, but a list will be found in the Forestry Commission *Catalogue of Publications* which is issued each year and can be obtained from the Publications Section, Forest Research Station, Alice Holt Lodge, Wrecclesham, Farnham, Surrey GU10 4LH.
(iv) In March 1989 the following publications, which were specifically concerned with arboriculture, were available.
    (1) *Bulletins*
        No. 65, *Advances in Practical Arboriculture*
    (2) *Arboricultural Leaflets*
        1. *The External Signs of Decay in Trees*
        2. *Honey Fungus*
        3. *Sooty Bark Disease of Sycamore*
        4. *Virus and Virus-like Diseases of Trees*
        5. *Common Decay Fungi in Broadleaved Trees*
        6. *Trees and Water*
        7. *Removal of Tree Stumps*
        8. *Phytophthora Diseases of Trees and Shrubs*
        9. *Verticillium Wilt*
       10. *Individual Tree Protection*

(3) *Arboriculture Research Notes*
These are issued by the Arboricultural Advisory and Information Service, and a list of those which are available will be found in the Forestry Commission *Catalogue of Publications* referred to above.
(4) *Miscellaneous*
Forestry Research Coordination Committee:
Report of the Review Group on Arboriculture (February 1988)

(*b*) *Arboricultural Association publications*
(i) In 1989 the undermentioned publications were available:
(1) *Handouts*
1. *Trees Suitable for Small Gardens*
2. *A Guide to Tree Planting*
3. *Young Tree Maintenance*
4. *Tree Management*
5. *Evergreen Hedges*
6. *Tree Roots*
(2) *Booklets*
A Guide to Tree Pruning
Trees on Development Sites
Trees on Golf Courses
Tree Survey and Inspection
Model Specification and Tender Documents
Evaluation Method for Amenity Trees
Evaluation Method for Amenity Woodlands
(ii) Further information regarding these publications can be obtained from the Secretary, Arboricultural Association, Ampfield House, Ampfield, Romsey, Hampshire SO51 9PA.

3 *Periodicals*

The following periodicals are either concerned with arboriculture or contain articles dealing with the subject or other closely related matters:

(*a*) *Arboricultural Journal*
 (i) Journal of the Arboricultural Association.
(ii) Published quarterly.
(iii) Editor's address: South Lodge, South Parks Road, Oxford OX1 3RF.

(*b*) *Forestry and British Timber*
 (i) Published monthly by Benn Publications plc.
(ii) Editor's address: Benn Publications plc, Sovereign Way, Tonbridge, Kent TN9 1RW.

(*c*) *The Garden*
 (i) Journal of the Royal Horticultural Society.
 (ii) Published monthly.
(iii) Editor's address: 80 Vincent Square, London SW1P 2PE.

(*d*) *Quarterly Journal of Forestry*
 (i) Journal of the Royal Forestry Society of England, Wales and Northern Ireland.
 (ii) Published quarterly.
(iii) Editor's address: 102 High Street, Tring, Herts. HP23 4AH.

(*e*) *Year Book of the International Dendrology Society*
 (i) Published annually.
(ii) Editor's address: Les Fontenelles, Forest, Guernsey, Channel Islands.

CHAPTER XXIV

Trees for Urban Areas

1 *General*

(*a*) This chapter contains a selection of trees which are suitable for planting in towns and urban areas. However, the word 'suitable' requires some qualification because a tree may be suitable under one set of conditions but not under another. In general terms, there are certain overriding qualities which a town tree should or should not have, and some of these are summarized below.

(i) Trees must never be potentially dangerous.

(ii) They must not be prone to cause damage to walls, drains or foundations, as in the case of poplars.

(iii) Trees that produce fruit which may be attractive to children, should be planted with considerable circumspection, since the seeds of some, notably the laburnum (*L. anagyroides*), are poisonous. Full details will be found in *Poisonous Plants and Fungi* by M. R. Cooper and A. W. Johnson (HM Stationery Office, 1988). Horse chestnut trees may be damaged on account of efforts by the young to dislodge the chestnuts or 'conkers'.

(iv) Broadly speaking, trees which grow to a very large size are not suitable in towns unless they are provided with all the space they need, as in parks and open spaces.

(v) Flowering trees may be very attractive during the time that they are in bloom, but this may be followed by a dull or unattractive appearance for the rest of the year.

(*b*) Owing to the limitations of space, it is possible to include only a small number of trees in this chapter. However, a great deal of valuable information will be found in *Decorative Trees* by A. F. Mitchell and J. Jobling (HMSO, 1984).

(*c*) In the following lists, an indication of the relative size of tree is given by the letters 'L' (large), 'M' (medium) and 'S' (small). The letters 'D' and 'E' are respectively used to denote deciduous or evergreen.

2 *Trees with attractive fruit*

Arbutus unedo – Strawberry tree (M/E)
  Fruit reddish-orange, somewhat strawberry-like
  19 mm ($\frac{3}{4}$ in.) diameter
Cornus mas – Cornelian cherry (S/D)
  Bright red plum-like fruit
  16 mm ($\frac{5}{8}$ in.) long
Crataegus crus-galli – Cockspur thorn (S/D)
  Fruit deep red
  13 mm ($\frac{1}{2}$ in.) diameter.
  Several species of *Crataegus* bear attractive fruit.
Ilex aquifolium – Holly (M/E)
  Red berries
  6 mm ($\frac{1}{4}$ in.) diameter
  Some species of *Ilex* have yellow berries.
Malus pumila – Flowering crab (S/D)
  'Golden hornet'
    Deep yellow fruit
    About 25 mm (1 in.) wide
  'John Downie'
    Bright orange and scarlet fruit
    32 mm ($1\frac{1}{4}$ in.) long
Sorbus aucuparia – Rowan or Mountain ash (S to M/D)
  Bright red berries
  Up to 10 mm ($\frac{3}{8}$ in.) diameter
Sorbus hupehensis – Hupeh rowan (S to M/D)
  Fruit white with a pink tinge
  9 mm ($\frac{5}{16}$ in.) diameter.

3 *Trees with attractive flowers*

Aesculus hippocastanum – Horse chestnut (L/D)
Cercis siliquastrum – Judas tree (S/D)
Crataegus oxycantha – May or Hawthorn (S/D)
Magnolia
  M. campbellii – Campbell's magnolia (L/D)
  M. denudata – Yulan (S to M/D)
  M. grandiflora – Bull Bay (L/E)
  M. x soulangiana (S to M/D)

Prunus spp.
  Prunus avium – Wild cherry or Gean (L/D)
  There are also a wide selection of Japanese cherries such as 'Amanogawa' and 'Kanzan' (S/D).
Sorbus
  S. aucuparia – Rowan or Mountain ash (S to M/D)
  S. intermedia – Swedish whitebeam (S to M/D)

4 *Trees with attractive bark*

Acer davidii – Père David's Maple
  Bark green to purple when young, becoming striped with white as age increases. (M to L/D)
Acer griseum – Paper-bark maple
  Bark peels in large flakes to show new orange-coloured bark beneath. (M/D)
Arbutus menziesii – Madrona
  Orange-red bark which flakes off to reveal yellowish-pink areas beneath. (M to L/E)
Betula jacquemontii – Jacquemont's birch
Betula utilis – Himalayan birch
  These two species are very closely related and both have very white smooth bark. (M/D)
Betula albo-sinensis – Red-barked birch
  The bark, which is orange to orange-red in colour, peels off in very thin sheets. (M to L/D)
Pinus bungeana – Lacebark pine
  The smooth greyish-green and brown bark peels off in flakes to show patches of green, cream, purple and brown. (L/E)
Prunus maackii – Manchurian cherry
  The bark of this species, which is smooth and shining, is brownish-yellow in colour. (S/D)
Prunus serrula – Tibetan cherry
  The striking, glossy brown, bark peels off horizontally in bands around the bole. (S/D)

5 *Trees for autumn colour*

Acer campestre – Common maple
  Yellow (S)
Acer japonicum – Japanese maple
  Crimson (S)

Acer nikoense – Nikko maple
  Red and yellow (M)
Acer platanoides – Norway maple
  Red, yellow and brown (L)
Aesculus hippocastanum – Horse chestnut
  Golden (L)
Amelanchier spp. – Snowy mespilus
  Red and yellow (S)
Betula spp. – Birch
  Yellow (M)
Ginko biloba – Maidenhair
  Golden yellow (L)
Liquidamber styriciflua – Sweet gum
  Scarlet, purple and orange (L)
Malus tschonoskii – Pillar apple
  Crimson, orange and purple (M)
Nyssa sylvatica – Tupelo
  Red and yellow (L)
Quercus coccinea – Scarlet oak
  Brilliant red (L).

6 Trees with a weeping or pendulous form

Betula pendula cv. 'Youngii' – Young's Weeping Birch (M/D)
Chamaecyparis nootkatensis cv. 'Pendula' – Weeping Nootka cypress
  (L/E)
Fagus sylvatica f. *pendula* – Weeping beech (L/D)
Fraxinus excelsior cv. 'Pendula' – Weeping ash (L/D)
Picea breweriana – Brewer spruce (L/E)
Picea smithiana – West Himalayan spruce (L/E)
Salix x chrysocoma – Weeping willow (M/D)
Tilia petiolaris – Silver pendent lime (L/D)

7 Trees with a fastigiate or columnar form

Calocedrus decurrens – Incense cedar (L/E)
Carpinus betulus cv. 'Columnaris' – Hornbeam (L/D)
Chamaecyparis lawsoniana cv. 'Kilmacurragh' – Lawson cypress
  (L/E)
Cupressus sempervirens var. *sempervirens* – Italian cypress (L/E)
Fagus sylvatica cv. 'Fastigiata' or 'Dawyck' – Dawyck beech (L/D)

Juniperus virginiana cv. 'Sky rocket' (S to M/E)
Populus nigra cv. 'Italica' – Lombardy poplar (L/D)
Prunus serrulata cv. 'Amanogawa' – Japanese cherry (S/D)
Quercus robur f. *fastigiata* – Cypress oak (L/D)
Taxus baccata cv. 'Fastigiata' – Florence Court or Irish yew (M/E)

8 *Trees for chalk soils*

Ailanthus altissima – Tree of Heaven (L/D)
Chamaecyparis lawsoniana – Lawson cypress (L/E)
Cupressocyparis x leylandii – Leyland cypress (L/E)
Fagus sylvatica – Beech (L/D)
Fraxinus excelsior – Ash (L/D)
Ilex aquifolium – Holly (M/E)
Pinus nigra var. *maritima* – Corsican pine (L/E)
Pinus nigra var. *nigra* – Austrian pine (L/E)
Populus canescens – Grey poplar (L/D)
Sorbus aria – Whitebeam (S to M/D)
Taxus baccata – Yew (S to M/E)

9 *Trees for moist damp sites*

Alnus spp. – Alders (M to L/D)
Crataegus oxycantha – Hawthorn or May (S/D)
Metasequoia glyptostroboides Dawn redwood (L/D)
Picea sitchensis – Sitka spruce (L/E)
Populus spp. – Poplars (L/D)
Salix spp. – Willows (S to L/D)
Sequoia sempervirens – Coastal redwood (L/E)
Sorbus aucuparia – Rowan or Mountain ash (S to M/D)
Taxodium distichum – Swamp cypress (L/D)

10 *Trees for dry or sandy soils*

Ailanthus altissima – Tree of Heaven (L/D)
Betula pendula – Silver birch (M/D)
Castanea sativa – Sweet or Spanish chestnut (L/D)
Gleditsia triacanthos f. *inermis* – Honey Locust (L/D). This form is thornless.

Pinus (L/E)
   P. muricata – Bishop pine
   P. nigra var. *maritima* – Corsican pine
   P. nigra var. *nigra* – Austrian pine
   P. pinaster – Maritime pine
   P. radiata – Monterey pine
   P. sylvestris – Scots pine
Robinia pseudacacia – Locust or False Acacia (M to L/D)
Sorbus spp. – Rowans, Whitebeam, etc. (S to M/D)

11 *Trees for seaside districts*

Acer pseudoplatanus – Sycamore (L/D)
Cupressocyparis leylandii – Leyland cypress (L/D)
Cupressus macrocarpa – Monterey cypress (L/E)
Griselinia littoralis – Kupuka (S to M/E)
Pinus (L/E)
   P. nigra var. *maritima* – Corsican pine
   P. nigra var. *nigra* – Austrian pine
   P. pinaster – Maritime pine
   P. radiata – Monterey pine
Populus alba – White poplar (L/D)
Populus canescens – Grey poplar (L/D)
Quercus ilex – Holm oak (L/E)
Sorbus aria – Whitebeam (S to M/D)

12 *Trees for shelter belts and windbreaks*

Acer platanoides – Norway maple (L/D)
Acer pseudoplatanus – Sycamore (L/D)
Carpinus betulus – Hornbeam (L/D)
Cupressocyparis leylandii – Leyland cypress (L/E)
Pinus nigra var. *nigra* – Austrian pine (L/E)
Populus alba –White poplar (L/D)
Populus canescens – Grey poplar (L/E)
Quercus ilex –Holm oak (L/E)
Tilia cordata – Small-leaved lime (L/E)

13 *Fast-growing trees*

Cupressocyparis leylandii – Leyland cypress (L/E)
Cupressus macrocarpa – Monterey cypress (L/E)

Eucalyptus gunnii – Cider gum (L/E)
Larix x eurolepsis – Dunkeld or Hybrid larch
Larix kaempferi – Japanese larch (L/D)
Nothofagus obliqua – Roble beech (L/D)
Nothofagus procera – Rauli (L/D)
Populus spp. – Poplars (L/D)
Pseudotsuga menziesii – Douglas fir (L/E)
Pterocarya x rehderiana – Hybrid Wing-nut (L/D)
Salix alba – White willow (M to L/D)

14 *Trees for street and roadside planting*

(a) *General*

When selecting trees for street and roadside planting the following points should be borne in mind:

(i) They should be of a species which is easy to establish and able to start growth quickly.

(ii) On the whole, they should tend to be of an upright habit rather than a spreading one.

(iii) They should not attain too large a size when mature.

(iv) They should not produce fruit or nuts which may encourage damage.

(v) They should be able to withstand a reasonable amount of pruning.

(vi) They should be robust, healthy and not prone to disease.

(vii) In narrow streets, since there is less space, small-crowned or fastigiate trees should be used. In wide streets larger-crowned trees can be planted.

(viii) It should be remembered that no matter how attractive a tree in blossom may look, the flowering period is relatively short.

(ix) Trees selected for roadside planting in rural areas should fit in with the pattern of the countryside and conform with the trees already growing in the vicinity.

(b) *Trees for street planting*
(i) *Narrow streets*

Betula pendula cv. 'Fastigiata' – Birch
Carpinus betulus cv. 'Columnaris' – Hornbeam
Gingko biloba – Maidenhair tree
Malus tschonowskii – Pillar apple
Prunus cerasifera cv. 'Pissardii' – Myrobalan
Prunus serulata cv. 'Amanogawa' – Japanese cherry

(ii) *Moderately wide streets*
Acer platanoides – Norway maple
Betula pendula – Silver birch
Crataegus spp. – Flowering thorns
Prunus spp. – Flowering cherries
Sorbus aucuparia – Rowan or Mountain Ash
Sorbus aria – Whitebeam
Sorbus intermedia – Swedish whitebeam

(iii) *Broad streets*
Aesculus hippocastanum cv. 'Baumannii' – Horse chestnut. This cultivar does not produce nuts or 'conkers'.
Carpinus betulus – Hornbeam
Platanus acerifolia – London plane
Tilia x euchlora – Caucasian lime.

(c) *Trees for rural roadside planting*
Aesculus hippocastanum – Horse chestnut
Carpinus betulus – Hornbeam
Fagus sylvatica –Beech
Fraxinus excelsior – Ash
Quercus petraea – Sessile oak
Quercus robur – Pedunculate oak
Tilia platyphyllos – Broadleaved lime

Note. The various species, varieties and cultivars which are mentioned in this chapter are only a selection of those which may be planted for the purposes which are mentioned. A much larger selection will be found in the appropriate books which are included in Chapter XXIII, Section 1.

Grants for Tree Planting

A number of grants can be obtained for planting trees in what may be termed an arboricultural context. These are available from certain government departments and other bodies in respect of small woods, groups of trees, shelter belts, individual trees and those which can make a contribution to nature conservation. Details are given below.

(*a*) *Grants for small woods*
(i) *The Countryside Commission*
    (1) Discretionary grants of up to 50 per cent of the approved costs are available for planting small areas of woodland:
        a. On sites of 0.25 hectares (0.6 acres) or less in open countryside.
        b. On sites of over 0.25 hectares if the scheme is not eligible for a grant from the Forestry Commission.
    (2) The species should normally be broadleaved and appropriate to the locality.
    (3) Further information can be obtained from:
        a. The Countryside Commission, John Dower House, Crescent Place, Cheltenham, Glos. GL50 3RA.
        b. The Countryside Commission for Scotland, Battleby, Redgorton, Perth PH1 3EW.

(ii) *The National Parks*
    (1) Assistance of up to 75 per cent of the approved cost is available for private landowners or occupiers who wish to plant bare land or rehabilitate broadleaved woodland within a National Park.
    (2) Such support may take the form of cash, a supply of trees and materials or a contribution by the applicant towards the cost of the work.
    (3) The area to be planted must not exceed 0.25 hectares (0.6 acres).

(4) Details are available from the appropriate National Park Authority, which can be located through the telephone directory for the area.

(b) *Grants for individual trees or small groups*
  (i) *Local authorities*
    (1) Some local authorities provide assistance for planting individual trees. For example, the Devon County Council will supply individuals, Parish Councils and other organizations with free trees subject to certain conditions as to the site, planting, protection and maintenance. Private individuals are limited to 25 trees in any one planting season.
    (2) Enquiries should be addressed to the appropriate local authority.

  (ii) *Ministry of Agriculture, Fisheries and Food*
    (1) Grants are available for planting, staking and protecting trees which are intended to provide shade for stock. These could be in hedgerows or on other approved sites.
    (2) Further information can be obtained from:
      a. The Agricultural Development and Advisory Service (ADAS) of the Ministry of Agriculture, Great Westminster House, Horseferry Road, London SW1P 2AE, or from the Divisional Offices of the Ministry (addresses can be traced through the local telephone directory).
      b. The Department of Agriculture and Fisheries for Scotland, Chesser House, 500 Gorgie Street, Edinburgh EH11 3AW.

  (iii) *The Tree Council*
    (1) The Council will consider providing a certain amount of limited financial help for planting trees.
    (2) Applications for assistance, which is limited to half of the total cost of planting, should be made to The Secretary, The Tree Council, 35 Belgrave Square, London SW1X 8QN.

(c) *Grants for shelter belts*
  (i) Grants are available for planting and protecting shelter belts and shelter hedges and for their maintenance during the first three years.
  (ii) Details can be obtained from the Agricultural Development and Advisory Service or from the Department of Agriculture for Scotland at the addresses given under Section (b)(ii), which deals with individual trees.

(*d*) *Grants from amenity planting*

(i) Assistance towards the cost of planting trees for amenity purposes (subject to certain conditions) may be obtained from the sources given below.

(ii) In some cases, planting for amenity may coincide with planting for other purposes, as for example, individual trees.

(iii) Enquiries should be made to:

The Agricultural Development and Advisory Service
The Department of Agriculture for Scotland
The Countryside Commissions
The National Parks
The Tree Council
Local Authorities

at the addresses which have already been given above.

(*e*) *Grants in relation to conservation*

(i) The Nature Conservancy Council will consider making grants for projects which it considers to be a worthwhile contribution to nature conservation.

(ii) As regards trees, the following would be considered:

(1) The planting of individual trees or small woods:

a. In areas designated as Sites of Special Scientific Interest.

b. On sites considered to be important wildlife habitats.

c. In locations which are managed as nature reserves.

d. On sites recorded in the Council's inventory of ancient woodlands.

(2) The management of existing trees or woodlands situated on any of the sites referred to above.

(iii) All grants are discretionary and usually amount to 50 per cent of the total cost.

(iv) Enquiries should be addressed to the nearest Regional Office of the Nature Conservancy, which can be obtained from the local telephone directory. Failing this, the Headquarters of the Council at Northminster House, Peterborough PE1 1UA, should be approached.

Index

In compiling this index, the following procedure has been adopted. Fungi, bacteria and insects are arranged in alphabetical order of English names. Species of trees will be found under the purpose for which they can be used or according to their special characteristics such as street planting, shelter belts, autumn colour or form. Books and manuals are not indexed individually but will be found under the section to which they apply or in Chapter XXIII. Periodicals are included under their titles which are printed in italics. The titles of statutes, statutory instruments and law cases are not included in this index.